INNOVATION
NATION

Canadian leadership from Java to Jurassic Park

INNOVATION
NATION

Canadian leadership from Java to Jurassic Park

LEONARD BRODY

WENDY CUKIER

KEN GRANT

MATT HOLLAND

CATHERINE MIDDLETON

DENISE SHORTT

WILEY
wiley.com

John Wiley & Sons Canada Ltd
22 Worcester Road
Etobicoke, Ontario
M9W 1L1

National Library of Canada Cataloguing in Publication

Innovation Nation : Canadian Leadership from Java to Jurassic Park / Leonard Brody .. [et al.].

ISBN 0-470-83202-9
1. Businesspeople—Canada—Biography. 2. Technological
innovations—Canada. 3. Entrepreneurship—Canada. 4. Success in
business—Canada. I. Brody, Leonard

HC112.5.A2I55 2002   658.4'21'092271   C2002-903677-1

Production Credits
Cover and text design: Sputnik Art + Design Inc., Toronto
Author photos: Natasha Nicholson Photography

Printed in Canada

10 9 8 7 6 5 4 3 2 1

CONTENTS

ACKNOWLEDGEMENTS

Innovation Nation was the result of an immense amount of hard work from a a team that selflessly donated their time, energy and talents.

Our sincere thanks to:

Bill Baker, Ron Bernbaum, Ian Burchett, Helen Burstyn, Ninon Charlebois, Kelvin Choi, Michael Corcoran, Rocco Delvecchio, Bernie Etzinger, Darlene Gibson, Felda Hardymon, Joseph Kay, Handol Kim, Claude Lajeunesse, Bobby LeBlanc, Michael Levine, Michael Lewkowitz, Loudon Owen, David Pecaut, Norma Pimentel, Rocco Rossi, Sheila Smail, Justyna Susla, and Pearl Yaffe.

A very special thanks to the team:

Matt Anestis, Karen Bray, Kevin Bright, Jamie Broadhurst, John Eckert, Richard Florizone, Joel Gladstone, Robert Harris, Val Jackson, Michael Lank, Jason Macdonnell, Elizabeth McCurdy, Karen Milner, Karen Satok, Ron Taylor, Martha Wilson and Valerie Ahwee.

We are also grateful to those profiled in this book for donating their time and thoughts; you are truly the heart of this great Canadian story.

This book would not have happened without:

Our creative director—Marcy Grossman—who developed the artistic vision, convinced us it would be unique, and then dedicated countless hours to make it happen. Her spark, humour, encouragement, and initiative were critical to seeing this project to the finish line.

DEDICATION Part of the vision of *Innovation Nation* was to highlight Canadian leadership in the technology economy, while at the same time, contributing to the creation of Canadian role models.

In this spirit, we would like to dedicate this book to each of our parents—our most valued role models of all.

FOREWORD

"THE REAL DIGITAL DECADE HAS JUST BEGUN." BILL GATES, CHAIRMAN, MICROSOFT FEBRUARY 2002

Central to Mr. Gates's view that the Digital Decade has just begun is the belief that information technology will provide over twice the productivity in the next 10 years that it did in the last. The other titans of the information age agree. Michael Dell likes to brag that Dell gets over an 800 percent return on its investments in IT due to the complete Web automation of its business. John Chambers of Cisco spends twice the amount on IT equipment that the average company does, and he swears this investment is why his company dominates the networking space. And if you think the next boom will be good only for the big guys, listen to what Amazon's CEO, Jeff Bezos, recently predicted: "For every successful startup that came out of the first wave of the Web, there will be 10 more started in the next wave."

It may seem unusual that the same writer who spent four years (1997 through 2000) warning people that 90 percent of all the Internet companies were ultimately going to go bust is now proclaiming that the Internet action—and upside—has just begun. I will go so far as to say that most of the entrepreneurial and investment opportunities associated with the Web are still ahead of us. Further, I believe that we do not even know the names of most of the companies that will eventually dominate the Internet age. If this seems hard to fathom, consider that the two largest PC manufacturers today—Compaq and Dell Computer—didn't even exist until after the PC stock bubble burst in 1983. It is also interesting to note that many of the biggest names in business, including Disney, Hewlett-Packard, Microsoft, and Cisco were started during recessions.

And it will be innovators, such as the Canadians profiled in this book, who will create the technologies and companies that will ride this huge wave. People like Tim Bray of Antarctica Systems Inc. and Rob Burgess of Macromedia who are working away as I write, inventing the next generation of the Web and its applications. It will be companies like RIM and Zero Knowledge that will show us how to seize on these new technologies and prosper and grow.

By the year 2005, two billion people will be continuously connected to a powerful global network of satellites and fibre optic cables. We will each carry a low-cost pocket Internet device that will serve as a daily organizer, e-mail device, telephone, radio, and video-playing machine, all rolled into one. Virtually every electronic device we own, from our computers to our television sets to our refrigerators, will be connected to the network and programmed to automatically communicate and work for us. Sony chairman Nobuyuki Idei recently summarized what this world would be like: "PCs and most consumer electronic devices will soon be connected seamlessly to one another over a variety of cable, satellite, phone, and wireless networks, and will provide users with instant access to almost all content and services from wherever they sit."

The second wave of the consumer Internet boom is already gaining significant momentum. Today more than 500,000 people depend on the Internet and the Web to communicate, perform their jobs, research and buy stocks, and get directions for driving. Canadian Internet users are savvy about the available options and have been quicker than most to adopt new applications of emerging technologies. These habits and technology-adoption rates bode well for the immediate future of the Web. Showing us the future is the emerging "Always-On" generation, a group of largely 12-year-old to college-age kids who use personal computers and handheld devices that are constantly hooked to the Internet. According to a study by the UCLA Centre for Communication Policy, the average kid spends 11.3 hours a week online. Over three-quarters of kids in a recent Pew survey claimed that the Internet plays a major role in their lives and that they would miss it if they could no longer go online. A whopping 94 percent use the Internet to research their papers, and 41 percent seek regular help from teachers and counselors via e-mail. So just as those of us in the PC generation led the last boom, these kids are the early adopters of new Web applications and services. They will show us how to appreciate text messaging and, very soon, visual messaging, as well as mobile commerce and other applications that will make their way into our everyday rituals.

The surge in the consumer Internet market has driven handheld and pocket device sales to new heights. The number of non-PCs connected to the Internet surpassed the number of connected PCs in 2002. In 2002 the cellphone user number passed the billion-user mark, crowning the fastest 10-year penetration of any product in commercial history. China alone is adding five million new cellphone users a month. By 2005 most cellphones will come standard with Internet access and a variety of Web service options. Text and visual messaging, already hugely popular in Japan and Europe, will emerge as the "killer application" for the next-generation Internet. According to the Pew report, 74 percent of online teens use instant messaging several times a week. AOL's president of advanced services, Ted Leonsis, reported that 1.2 billion instant messages were sent over the AOL network on September 11, which underscores our increased dependence on the online world as a way to communicate. The number of annual text and visual messages sent over the Internet is expected to reach a whopping 1.2 trillion by 2005. Canada continues to play a leading role in the evolution of technologies like this. That much is clear from the evidence of the past few years.

But we shouldn't let all this talk about the mobile device boom overshadow what is happening in the larger device market. Just when we felt the PC had become a horribly boring commodity device, Apple Computer CEO Steve Jobs raised his voice (and his new iMac) at MacWorld 2002 in San Francisco and proclaimed the beginning of the third wave in personal computing. In the new era, Mr. Jobs sees the desktop machine as a digital hub for the home that connects our video and still-picture cameras, as well as our CD and DVD players. Across the Pacific, Sony's Mr. Idei will give you a slightly different spin on the same vision. While Sony also produces stylish PCs whose robust sales have been resisting the PC slump, the company is a leader in television sets as well. "We think consumers will use their TV like a server to download and manage most of their entertainment audio and video content, because TVs will always serve the captive audience," Mr. Idei says. To facilitate this, Sony has developed a prototype product called the Personal Network Home Storage System, which can store up to 450 hours of DVD movie content, 1,500 CDs, and 600,000 high-resolution images. Using a wireless home network, consumers can use their TVs to manage and interact with their Walkmans, PlayStations, and video cameras.

Apple and Sony are not the only pioneers in the much-hyped world of converging consumer electronics. At the recent International Consumer Electronics Show in Las Vegas, Moxi Digital, a Silicon Valley startup, and Microsoft have both jumped into the game. Moxi's digital TV set-top box with a high-speed Internet connection is also designed to function as a home media server and integrate the functions of a consumer's digital cameras and media players. The company, started by WebTV Networks cofounder Steve Perlman, intends to license its system to cable and satellite TV operators. Microsoft, for its part, designed its new Xbox game machine to serve also as a digital hub that can download and store music CDs and DVDs. And their new operating system, Windows XP, has been designed so a user's PC experience can be easily extended over a wireless connection to multiple devices and smart monitors around the home. "I believe in basically 90 percent of Steve Jobs's vision for the future of the PC," Bill Gates told me at this year's World Economic Forum in New York.

On the business-to-business side as well, most of the Internet innovations are still to come. In the year 2000, there was $131 billion (U.S.) worth of business-to-business e-commerce. Depending upon which research firm you want to believe, this number is expected to grow to somewhere between $4 trillion and $7 trillion by the year 2005. Fueling this growth is an industry-wide push to create what is fashionably being called in Silicon Valley a "real-time" computing environment, where businesses use the Internet to automate their entire business processes. A key new Internet standard called XML will be at the centre of this changed environment, providing an artificial intelligence capacity to what has been a fairly static platform. XML is one of the many reasons Canada has such a solid claim on its "Innovation Nation" status. It's this kind of change that is powering the Web forward right now and Canada has continued to lead the way.

Large and small companies will continue to leverage these technologies to automate their businesses entirely. Trillions of dollars in transactions will flow spontaneously over the Internet, 24 hours a day, without human assistance. Cisco CEO John Chambers recently boasted that his company practices what it preaches, and has emerged as the largest e-commerce company on the planet, with well over 90 percent of their $20 billion in annual sales in some way facilitated over the Internet. Michael Dell also claims that over 50 percent of his orders come directly over the Internet.

The most advanced companies, like Cisco Systems and Dell Computer, achieve this productivity using smart networks that augment human intelligence by analyzing and organizing material, and automatically answering questions and alerting people when interesting things happen. These new Web functions are being made possible in part by that new-generation Web language, extensible markup language (XML). In essence, XML and its associated developments are making Web development a lot more efficient and much, much cheaper. To build a next-generation Web site, developers will be using software code like building blocks that are interchangeable and easily snap together. Many of these building blocks will be made from open-source software available on the Web, often at no cost. XML also allows developers to code different forms of Web content, like video, pictures, and text, and link them to dictionaries and encyclopedias that help interpret and sort this content for the user automatically. Microsoft CEO Steve Ballmer believes that "the emergence of XML as the *lingua franca* of the Web is the next big thing in computing, and we are betting 100 percent of our strategy on that change."

Perhaps not surprisingly, the inventor of the World Wide Web, Tim Berners-Lee, is at the centre of the creation of the next Web, which he refers to as the "Semantic Web." "The Semantic Web will be more intuitive, will understand the meaning of words and concepts, and will do so automatically," says Mr. Berners-Lee. (For more information on the Semantic Web, see www.semanticweb.org.) In this new environment, consumers will be able to dispatch intelligent software agents to cruise the Web, watching out for releases from their favourite artists, or buying items when they are available at a targeted price. Business will use these agents to seek out and even negotiate with suppliers for the cheapest prices, and then automatically procure sophisticated transactions, feeding them back into a company's financial and inventory systems.

Mr. Berners-Lee's vision is not lost to the entrepreneurial community. "The adoption of network-enabled applications by businesses is the single most important factor in future productivity growth," said Cisco Systems president and CEO, John Chambers, at a recent Salomon Smith Barney media conference. He went on to boldly predict that these smart networks will increase U.S. productivity rates over the next 10 years by almost double the current U.S. government estimates. And for those in the business of selling technology applications, perhaps the most encouraging part of Mr. Chambers's presentation was that most companies are not even close to adopting the next generation of the Web. Barely three percent of U.S. businesses have completed more than 50 percent of Wave 1 and 2 applications, according to a private Cisco-commissioned study.

The new Web will become so central to the way all companies interact that virtually all existing businesses will have to completely rewrite their Web sites in the new *lingua franca* to remain competitive. Keith Fox, the longtime vice president of corporate marketing at Cisco, which is just finishing a major rebuild of its Web site, is such a believer that he sees the next Web as the primary brand execution vehicle. He notes that if your customers, suppliers, and employees are increasingly transacting and servicing each one another over the Web, then it is important to consolidate and manage your brand in that environment. The new Web "allows CEOs to simply alter their messaging for the entire universe they serve," says Mr. Fox.

Zooming back up to 30,000 feet, we can now see that the promise of the information age, which was initiated back in the 1950s by the invention of the transistor, is finally being delivered to our global society at large. Information technology and the Internet are completing our transition from an economy of the land to an economy of technology and science. As Israeli Foreign Minister and Nobel Peace Prize winner Shimon Peres expressed to a gathering of city mayors in Rome last spring, "In this new economy, it is not the size of the land but the level of knowledge that will create new opportunities." The level to which technology has penetrated a nation's population will determine its ability to become rich or remain poor. Mr. Peres uses Japan, a country with a relatively small geography and limited natural resources as an example. "The only thing they really have is that they are Japanese," he said. "Look at the opportunities you can create with Japan today! More than you could with a country full of oil, silver, or gold."

Canadian innovators and entrepreneurs have instinctively and intuitively grasped the importance of that insight. Over and over, the people profiled in this book have stressed the importance of working with top people—recruiting them, training them, and keeping them. These lessons will continue to be hugely important for Canada as more and more of its snowbird entrepreneurs come back home.

In a world "globalized" by digital networks, the power and effectiveness of national governments are also fading. With the ability to access almost any information from any location, the new world citizen will become increasingly empowered and independent. The rise of the al-Qaeda terrorist network also revealed how the Internet can create cyber nations uninhibited by geographical boundaries or limitations. In this new age, the creators and distributors of information, such as the major media outlets, will also become more powerful than central governments. As a result, local institutions and the private sector, which serve their constituencies "eye-to-eye," will have to take more social responsibility.

To secure our future in a new networked world, where a small group of terrorists can communicate from remote locations and initiate mass destruction, we must be prepared to give up some of our privacy. To monitor its safety, the city of Washington, D.C., has already installed small digital cameras at most traffic intersections and public parks. Over time, we will be required to identify ourselves biometrically, perhaps by fingerprints, hand prints, or retina scans. We must also keep a keen eye on the hacker world. Today, most hackers are in their teens and are primarily showing off their hacker skills, rather than stealing credit card numbers or damaging information systems with viruses. But it only takes a few to incite havoc and inflict serious damage.

In the end, however, it is my belief that we will end up in a more secure and private world. Historically, whenever a tension point is created in the world, innovative entrepreneurs jump in to create solutions to ease the tension. We must appreciate and encourage the entrepreneur, and foster Silicon Valley-style entrepreneurial capitalism around the globe. It is our view at *Red Herring*, due in part to the instant communications power of the Web, that technology, scientific innovation, and entrepreneurship are spreading rapidly around the world. Europe, Japan, and South Korea are leading the wireless booms. Biotechnology innovation is showing up in places like Singapore and Cuba.

The bottom line is that no matter which consumer or business-to-business Internet usage numbers you look at, everything is growing at exponential rates. The only debate going on now is whether we are going to grow at a rate of 10 squared or 10 to the eighth. This book profiles some of the Canadian entrepreneurs and companies creating the new technologies which are inspiring us to welcome the global digital network. Their innovations will radically change personal and professional lives. In just the next few years, virtually everybody and everything will be instantly accessible. I expect, as has been the case in the past, that Canada will lead us into the adoption and growth curve of this exciting place in history.

THE PRECIS

This book sets out what many consider to be a well-kept secret. This secret has become an essential element in the competitiveness and profitability of many companies. It can make business builders more successful, investors wealthier, policy-makers more effective at encouraging economic growth, and may even embolden the leap into entrepreneurship. It is this: **Canada is a global technology leader—it is an *Innovation Nation*.**

From the invention of Java (the groundbreaking software that powers business computers), to developing the 3-D animation software used in *Jurassic Park* (which dramatically changed the way Hollywood made movies), Canadians have slowly moved toward pole position in the technology economy. Canadians like Jim Balsillie and Mike Lazaridis championed the idea of wireless e-mail and built Research in Motion into a Canadian-based global success, revolutionizing the way companies communicate. Jeff Skoll, the Montrealer who co-founded the online auction company eBay, successfully competed in the U.S., arguably the toughest technology market in the world. Vancouverite Tim Bray is working alongside Web creator Tim Berners-Lee at the heart of the World Wide Web Consortium to re-engineer the Internet into its next phase. Michael Potter, the founder of Cognos—one of Canada's most successful software companies, and Tom Jenkins—the innovator behind Open Text—have shown the ability of Canadian software product and business models to withstand even the toughest recessionary periods. Across the spectrum of technology, Canadians have either assisted or quarterbacked some of the most meaningful innovations of our time.

The notion that Canadians are world-class technologists is not particularly top-of-mind when defining the Canadian identity. For those outside Canada, particularly Canadians who left home many years ago because the country was burdened with punitive tax rates and an economy that did not celebrate wealth creation, the idea of Canada as an Innovation Nation is even more foreign. Times, though, have changed.

The Canadian technology sector has been diligently writing a new chapter through-out the last decade. It has, however, been quiet relative to the few Canadian stories of operatic proportion—Nortel and JDS Uniphase. Moreover, it is often deafened by the overtures of success coming from the United States. In addition, our underplayed achievements often come as a result of the fact that many Canadian entrepreneurs chose to sell their successful companies to larger organizations, a less headline-worthy exit than raising capital through an IPO on the public markets. Unfortunately, this general lack of awareness is costly in the face of increased competition for global investment dollars. Without our story being touted, our top talent is susceptible to recruitment to other geographies more clearly defined as "where the action is." At home, public policy-making aimed at encouraging growth and expansion will not be as effective if it's based on outdated ideas of reality.

This book was written to produce a record of Canadian accomplishment; to many of us, the profiles contained in it are energizing and a source of great pride. For several of the individuals featured here, their stories conclude with great wealth being created; for many others, with great wealth being lost. Throughout the pages to come, however, you will meet some of the minds who are the pillars behind our current success and sustained future growth. Many have played multiple roles in this grand play. For simplicity's sake, the actors can be divided into five essential groups. They are:

- *Serial Entrepreneurs.* Serial entrepreneurs are individuals who have launched a series of companies. They are a measure of maturity in an economy. Each business-building attempt, success or failure, adds to their arsenal of experience, such as lessons about defining opportunity, finding markets, securing funding, designing their organizations, and networking. Scars and victories create a basis for judgment and capability in committed, entrepreneurial "lifers." The lessons of the past contribute to raising the odds of success in the future. Canadians now have enough world-class experience that our nation's presence in the technology market will most certainly be accelerated over the next decade.

- *Financiers.* Most entrepreneurs have a very close relationship with, and rely heavily on, their investors. "Smart money" is aggressively sought to create synergies, thought partnership, and guidance. While seeing eye-to-eye can be challenging, mutual respect drives a bond that is productive in growing companies through each phase of their natural lives. Canada's financiers have come of age, providing entrepreneurs with a wide array of investment sources, led by an experienced venture capital and angel community. As Paul Chen, founder of FloNetwork, noted about his venture capital firm, McLean Watson Capital, they were critical partners with him in surviving the "long walk through the night" as his business searched to reinvent itself. Many of the entrepreneurs in this book have become financiers as well, creating a powerful hybrid of knowledge and access to capital.

- *Coaches.* Part of the inherent value for a nation in producing serial entrepreneurs is their ability to teach. Entrepreneurs who are willing to give back, through coaching and mentoring, have traditionally been the hallmark of evolution in the leading technology nations of the world. The U.S., in particular, has been very successful at rallying entrepreneurs to assist in developing a strong community. Today, Canada too is reaping some of the benefits of its entrepreneurs' prior successes. Many of the people profiled in this book have taken on the role of a coach in one way or another. They come in many shapes and locations—from founder of Onvia.com Glenn Ballman's Shawinigan Lake Club, to co-founder of the DocSpace Company Sandra Wear's Tykra, a consulting firm that guides entrepreneurs through hypergrowth. Though there are often differing levels of commitment—some formal, some informal—each has taken responsibility in ensuring that Canadian companies have a distinct experiential edge.

- *Academic Bridges.* Canada's post-secondary institutions are the key drivers in producing highly capable engineers, researchers, and scientists who make up much of the technological elite of this country. They are also an unparalleled source of ideas that can be commercialized into actual companies.

- *Infrastructure Builders.* Entrepreneurs are necessary but not sufficient in building a technology economy. The pre-condition required for accelerating innovation is a solid and stable infrastructure. Canada has one of the best technological infrastructures in the world. Creating this backbone has been one of our greatest strengths as a nation. Those who took part in building our infrastructure have played a critical role in paving the way forward.

These players, when looked at collectively, create a picture that resembles the early maps of Canada, with some territory known but much of it a vast expanse of possibility waiting to be explored. Like any early map, this one will be incomplete, and in the not too distant future, will look outdated. But like all good maps, an early outline is established for others to explore.

THE BOTTOM LINE—CANADA IS STRONG AND GETTING STRONGER

Canada's impact on the technology economy has been far out of proportion to the nation's modest size. Canadians have founded or directed some of the most revolutionary and talked-about companies of the new economy, including such names as Inktomi, Akamai, Red Hat, and Macromedia. Canadians have also risen to the top of the largest technology companies in the United States—for example, providing a president and COO to Yahoo in Jeff Mallet.

Canada has consolidated its platform for accelerated growth—the country now has a large and growing number of technology entrepreneurs, sophisticated and experienced venture capitalists, and other capital providers, as well as coaches and mentors. Postsecondary institutions are increasingly focused on the power of leveraging their academic intellectual goldmine in partnership with entrepreneurs and financiers.

Together these players have learned to combine their skills and capabilities to create technological innovation. These communities of technology innovation—the clusters—are typically anchored/linked with an academic institution, and a large technology-oriented company or two. Financial partners with deep sector knowledge are often part of the community, as are mentors such as entrepreneurs-in-residence. Canada has mature clusters forming within virtually every major community across Canada.

Economists have long understood the importance of innovation to a country's productivity and, ultimately, to its improved standard of living. Our governments are beginning to understand this; our academic institutions are promoting it, and our businesses are looking to innovation to keep them competitive in an increasingly global world. Innovation is now widely seen as the key to building a stronger Canada. Toward that goal, all players are aligning practices and policies to nurture an environment conducive to growth.

At the same time, the broader economic context is moving more and more in our favour. The public sector has made substantial strides in reshaping the tax and regulatory environment to encourage business-building in Canada. The capital gains rate has been cut and is actually on par with, and often lower than, the rates south of the border. Corporate tax rates have been reduced; by 2005, Canadian firms will enjoy corporate tax rates lower than those in the United States.[1] In addition, Canada's public stock exchanges are working through a restructuring that is making them friendlier to entrepreneurs. This, combined with Canada's remarkable cost advantage for doing business, makes this country exceptionally attractive to the global business community.

Canada has the foundations in place, and is poised for an even better showing in the years to come.

THE DNA OF *INNOVATION NATION*

We have organized this book into four distinct sections.

1. *Setting the Context (the Foreword):* Anthony Perkins is the editor-in-chief and founder of *Red Herring*—the industry bible for technology entrepreneurs. Early on, we contacted Anthony and told him that the Canadian story needed to be told in his publication. He replied, "What Canadian story?" He has been a believer ever since. In the foreword, Tony confidently sketches a future where technology becomes more and more intertwined with our daily lives. He also sheds some light on a new paradigm for entrepreneurs to think about when building new enterprises.

2. *Accelerating Opportunity (the Precis, the Introduction):* We discuss in detail the underpinnings behind Canada's stature as a technological leader and why we believe we have such a strong story to tell. We assume the perspective of an investor, and map out an argument about why Canada is a good bet these days, and why the odds may be getting better. We also take a stab at outlining how this country can benefit investors, policy-makers, and budding entrepreneurs.

---

[1] Canadian Minister of Industry's Keynote address to the Committee for Economic Development and eBusiness Roundtable on May 16, 2001. http://www.ic.gc.ca/cmb/Welcomeic.nsf/503cec39324f7372852564820068b211/85256a220056c2a485256a8f00561171!OpenDocument

3. *The Stories (30 Profiles):* Thirty Canadians have been selected for profiling. They are drawn from a broad array of organizations, large and small, from Canada and the U.S. One of the features of the people in the profiles is the fluidity of their careers—many will not even be in the same office by the time this book is printed. The true success of this project will be demonstrated when, while leafing through this book, a reader says, "I didn't know a Canadian did that."

4. *The Truth about Entrepreneurship (the Afterword):* Dr. Paul Kedrosky was, until this year, a professor at the University of British Columbia and ran one of North America's only university venture capital programs. Paul lives something of a dual existence as a Canadian, living and working in the U.S. but teaching in Canada and mentoring many startups in this country. He is a partner in a U.S. venture capital fund and a regular contributor to the *Wall Street Journal* and *Wired*. Paul's piece challenges the reader with the truth of what life is like as a technology entrepreneur. He provides a context from which to measure and appreciate the accomplishments of the Canadians featured here.

THE MAKING OF *INNOVATION NATION*

This book is the legacy of an extraordinary three-year collaboration between the public (Industry Canada/Investment Partnerships Canada) and private sectors that came together to form the Canadian E-business Opportunities Roundtable. The Roundtable functioned as an advisory group to the nation on the state of technology and innovation north of the forty-ninth parallel. The mission was an ambitious one: to co-operatively re-engineer Canada toward e-business leadership in the global economy. The challenge was to collect the best in the country on an ad hoc basis with every participant agreeing to donate the asset he or she had the least of—time.

Chaired by David Pecaut of the iFormation Group and John Roth, the former CEO of Nortel, the Roundtable's membership included leaders from federal and provincial governments and 35 of the top executives from the Canadian technology sector and academic institutions.

Annually, the Roundtable released a report card that graded the country's development. Each year, objectives were set and missions were undertaken. Through this process, the Roundtable essentially functioned as a rudder of an overarching movement that assisted in turning the Canadian economy around—a movement whose success can be measured in such accomplishments as helping build the case, and lobbying for, the substantial lowering of the country's taxes; encouraging Internet adoption by Canada's small businesses to improve their productivity; the furtherance of the innovation agenda; and, in some respects, this book.

Most of us met through the Branding Team, one of the five working teams of the Roundtable. Those 20 individuals were challenged to spread the word about Canada as a technology leader. The truth was so compelling that we decided to collect the best leaders and entrepreneurs Canada had to offer and let them tell the story for us. There was no spin necessary at all. The story is at least as strong, if not stronger, as other countries (Israel, Ireland, India), which appear to have higher public profiles. We decided a book would be a fitting and lasting testament to the achievements our nation had attained.

WHAT THIS BOOK IS, AND WHAT IT IS NOT

The intention of this project was not to write a treatise on innovation in Canada. Nor was it the goal to outline a history of the technology sector in this country. More importantly, this book was not written to pass judgment. Some of the entrepreneurs featured in this book chose to establish their companies outside of Canada; some could not have done it in Canada even if they wanted to. Regardless, we decided not to pontificate as to whether a Canadian-based entrepreneur is somehow more Canadian than one who left. Common to many of the individuals we interviewed here is the theme of border transparency, the state of mind that requires Canadians to adopt a North American mindset in order to succeed in their business. The point of this book is not just that Canada is a great place for entrepreneurs; it is equally the case that we produce world-class entrepreneurs, wherever they reside.

Each person profiled in this project has experienced both the height of success and the valley of failure. That is why they are interesting. You may read through the pages ahead and think, "why is this person considered an innovator?" The answer for each one is the same. They all contributed to changing the landscape of the Canadian or global technology scene forever.

This book is a celebration. It is no secret that Canadians have been remiss in telling their story. We are not known for our boasting. We hope this project will help Canadians to understand how far we have come, and how bright the future looks.

Another reason for writing this book was to inspire and nourish our young entrepreneurs. Therefore, the net royalties for this project will go toward an ongoing prize at Ryerson University for the best business plan in the IT Department's e-Business program.

Finally, we would like to thank the members of the Branding Team, and all those who supported the development of this book. We are extremely proud of the result. It was not as hard as we thought in many ways, but much harder in ways that we never would have imagined.

INTRODUCTION

Over the last decade, Canada has gone through a dramatic metamorphosis. It is almost unrecognizable from its former self (a nation that, not long ago, was written off by the foreign financial community as hopeless). The "New Canada," has seen remarkable progress on technology-friendly tax policies—with capital gains, and soon corporate taxes, dropping to rates equal to, or even lower than, those in the U.S. The "New Canada" encourages trade and helps local companies access foreign markets. The "New Canada" also provides opportunities for Canadian entrepreneurs to build their businesses at home by leveraging new high-tech clusters with experienced serial entrepreneurs and a venture capital community that is finally coming of age. The "New Canada" became a launch pad for international companies looking to enter the U.S. market. The removal of barriers for foreign investors to invest in Canada, particularly large international capital pools intersted in investments, has made Canada much more friendly to the flow of capital and ideas into Canada.

Like a highly effective prizefighter, Canada is beginning to punch well above its weight when it comes to technological innovation. Table 1 below lists how Canada has performed in a number of recent studies on the matter. When reading them, remember that Canada's population was just some 30 million in 2001—about one-tenth that of the United States, half that of the next smallest G8 member, or about thirty-fifth in world population size.

TABLE 1 – CANADA'S INNOVATION AND "HIGH-TECH" RANKINGS

- Number three in the Global Competitiveness Survey published by the Harvard Centre for International Development (2001).[1]

- Number one in the world in individuals achieving college or university education.[2]

- Eight of the world's top 100 business schools are in Canada, according to the U.K.'s *Financial Times* 2002 rankings (more than any country except the U.K. and U.S.).[3]

- Eighteen of North America's top 40 electrical engineering schools are in Canada (Gourman 1998).[4]

- Number four in "e-readiness" (Economist Intelligence Unit and Pyramid Research, 2001).[5]

- Highest or near-highest in the world in penetration and use of the telephone, cable, TV, computers in the home, and Internet access, especially high-speed access.

Where has this success come from? We believe that it stems from a recent maturity within Canada, as the people—researchers, entrepreneurs, investors, engineers, technologists, mentors, and larger companies—involved in the innovation community deepen and expand it. Industry also recognises the importance of this development and has been collaborating with the public sector to encourage these communities of ideas and talent, strengthening the country's fibre. These networks are not parochial, but extend across Canada, into the United States, and further afield, to international markets.

[1] Centre for International Development at Harvard University. Global Competitiveness Survey.
http://www.cid.harvard.edu/cr/pdf/GCR0102%20Overall%20Rankings.pdf
[2] The Government of Canada. Knowledge Matters: Canada's Innovation Strategy (2002). www.hrdc-drhc.gc.ca/stratpol/sl-ca/doc/summary.shtml
[3] *Financial Times*. 2002 MBA Rankings. http://specials.ft.com/businesseducation/FT3S5ND9MWC.html
[4] "The Gourman report." Dr. Jack Gourman—Undergraduate programs, 10th Edition 1998—as cited in http://www3.telus.net/info/ee-prog.htm
[5] The Economist Intelligence Unit. 2001 e-readiness Rankings. http://www.ebusinessforum.com/index.asp?layout=rich_story&doc_id=367

The second important driver of change, as mentioned above, has been learning how to manage this environment of technology entrepreneurship—meaning tax regimes, education, market access, technology infrastructure, and support for clusters. Substantial strides have been taken to manage these pre-conditions for success, building a stronger platform for these maturing networks to leverage.

OUR ROOTS

Canada has had a historical bond and national preoccupation with technology. Technological innovation was often adopted at a much faster pace in this country than was the case with our neighbour to the south. In the name of nation-building, Canada deployed technology to overcome its geographic handicap and bind the country together. The effort always carried with it a cultural imperative, as we have struggled to maintain a separate identity from the United States. One of the first examples of this was the public-private partnership of the late 1800s to use new steam engine technology to unite the country with a railway from coast to coast. While commerce was certainly a key objective, the more abstract, yet compelling, objective of uniting the country gave the construction of the Canadian railway an idiosyncratic and uniquely Canadian flavour. It became a format that would be pursued in this country for years to follow.

In the twentieth century, government and business continued working together in the same spirit to create a system of new enterprises using communication technologies—from the Canadian Broadcasting Corporation to Northern Telecom. The vision was not different from that which built Canada's railway. If technology could unite the country, then it would be implemented to do so. Enabling Canadian voices, in all senses of the word, was a strategy that was starting to work and becoming a part of our national fabric.

From these beginnings, the national objective broadened to developing Canadian companies that could compete on a global scale. Through government procurement practices targeted to build a critical mass of demand, favourable research and development tax policies, a system of government-sponsored laboratories and ongoing support for engineering and technical training, the federal and provincial governments encouraged the growth of technology firms like Systemhouse, Mitel, Gandalf, and others. In addition, commitments were made to attract global players such as IBM to undertake more of their research and development in Canada.

Throughout these collaborations, the challenge has been to maintain a balance between protectionism to nurture business development and the freedom to compete. With the negotiation and passage of NAFTA in the early 1990s, it was clear that the commercial agenda for the country was now focused on competing successfully both at home and internationally.

SERIAL ENTREPRENEURS—THE NEO COUREURS DU BOIS

When one thinks of entrepreneurs, images arise of the dedicated evangelist working in solitude to invent, and then develop, at great personal cost, the insights necessary to reap eventual payback. Canada has a rich historical metaphor for this—the "coureurs du bois" which translated to english means, "runner of the words." These individuals were early adventurers who explored and built businesses in regions of this country long before most Europeans would dare to visit. They passionately traversed our wilderness in pursuit of economic gain. The coureur du bois relied on his wits, worked extremely hard, traveled light, and was essential in opening up Canada for succeeding generations. They were restless, comfortable with solitude and, indeed, adversity only increased the value of their efforts.

An archetype of the serial entrepreneur is Terry Matthews, chairman and CEO of March Networks Corporation. He has arguably built more wealth in Canada, directly and indirectly, than any other entrepreneur. He co-founded Mitel (*Mi*ke and *Te*rry's *L*awnmowers) with Michael Cowpland in 1972 and grew the telecommunications hardware manufacturer to a leading PBX and key system provider that was sold to British Telecom in 1985. He went on to found Newbridge Networks in 1986 and subsequently sold it to Alcatel for $7 billion. Matthews reacquired the communications network division of Mitel. Merging Mitel's division with his newest company, March Networks,

Matthews plans to provide hardware and software support voice, data, and video high-speed networks. Matthews is also the founder and principal investor in Celtic House International, an early-stage technology venture capital firm with offices in Canada, the United Kingdom and the United States.

While our understanding of technology innovators is evolving, so too, is our knowledge of how they interact with complex environments. What emerges from the academic literature, and from these profiles, is a picture of serial entrepreneurs as individuals with great creative and analytical abilities, resilience, and a willingness to accept apparent contradictions—and pragmatism, not fatalism, about failure.

Many of the individuals profiled in this book speak about creativity and imaginative thought, and the ability to reflect on patterns they observe in a particular landscape. This innate creativity even appears to be part of our entrepreneurial identity as a nation. It is a source of national pride that Canadians make up a sizeable portion of the comedic pool south of the border (Mike Myers, Dan Ackroyd, and Jim Carey, to name a few). Jesse Rasch, founder of Inquent and Aprilis Ventures, attributes his success to his three Canadian genes of, "persistence, creativity and salesmanship." With those three attributes, he argues, "you can do just about anything." Gavriel State and Vikas Gupta created TransGaming Technologies to bring games to the Linux operating platform. Gupta believes that to be successful requires Canadians to go beyond the realm of creativity and to push aggressively to another level—a higher state. Says Gupta, "My favorite terminology is what I call the Wow factor. Everything that you do has to be representative of the Wow factor, meaning that if you do something, your employees, your clients, your business relationships, all have to turn around and say 'Wow! That's really cool. That's really innovative.'"

Resilience is a characteristic that appears in almost all of the innovators we spoke with during this project. Most of the people included in the profiles have suffered severe setbacks at some point in their careers, some even losing their businesses entirely. Why would we include innovators who have recently had a failure in a book about innovation? Because failure is a motherlode of experience to mine, and for serial business builders, learning from a failure can be brought forward into the next venture. Venture Capitalists such as John Eckert from McLean Watson Capital, "would rather invest in an entrepreneur who has fallen flat on their face, yet returns with another good idea looking for money. These entrepreneurs have been tested, and have shown enormous resilience in the face of failure, and are often times, a much better bet to back than those who have yet to experience failure."

James Fierro is a classic example of the highs and lows of technology entrepreneurship. Fierro currently heads a Vancouver-based investment and startup consulting firm called the Venture Resource Group and has become an angel investor leading many successful startups. Fierro founded Infotag Systems Inc., which designed, manufactured, and sold Radio Frequency Identification Systems. He raised almost $5 million prior to taking the company public in 1995. He then co-founded and provided the financing behind HomeGrocer.com, the first company to provide fully integrated grocery shopping online. The company raised $20 million in its initial IPO, and eventually merged with Webvan. The new partnership failed and became a highly publicized early casualty of the dot-com experience. Despite this, lessons in hand, Fierro moved on to his next startup ventures, including Recipco, an Internet communication and transaction company, and COL.com, an Internet marketer of tourism products. Both of them appear to be on the runway toward success. As Fierro saw first hand, innovation and resilience are the same thing.

Family plays a large part in the success of many serial entrepreneurs. With parents who worked at such jobs as self-employed real estate agents, farmers, and inventors, many of those profiled spoke about growing up with the risks associated with a somewhat unreliable income stream. As Hammie Hill noted, "One of the fundamental factors in being an entrepreneur is having massive risk tolerance. Growing up seeing the ups and downs, financially, that we went through as a family, we knew that the world doesn't end if things are tight for a while."

The ongoing commitment to building businesses, and the extension into mentoring and financing other entrepreneurs, is a pattern that seems to have established itself as a foundation of the business cycle in this country. The circle is completed, after success or failure, when the serial entrepreneur spends time passing his or her wisdom to the next generation. As Evan Chrapko, founder of the DocSpace Company, points out, "If half of one of my words lights up some kind of spark or fire in a fisherman's son, or an Inuit daughter, or someone else sitting in a tractor cab in the middle of nowhere, who doesn't know whether they have the right to think that they can do it too, I need them to know that they can." A chain reaction has clearly begun. Canadian entrepreneurs finally have heroes—the Neo Coureurs du Bois.

FINANCIERS AND COACHES—THE EXTENDED FAMILY

While it is tempting to embrace the image of the serial entrepreneur as romantic genius, a growing body of research shows that working away at solving a seemingly insoluble puzzle, to be an infrequent, almost rare phenomenon. Entrepreneurs interact with a complex environment to develop their ideas, secure early financing, recruit technical and managerial talent, go to market, and move to the next stage of business development. One of the most critical partners that an entrepreneur can have in this process is a financier.

Not long ago, technology innovators in this country had a limited choice of capital providers—typically they turned to the banks once their credit cards were maxed out and the house was mortgaged for the second time. Unfortunately, an entrepreneur with a good technology idea but no assets to secure the loan was faced with a very challenging environment in which to raise money.

In the 1980s some of the large Canadian capital pools—typically pension funds and other money managers—began to take tentative steps into the world of venture capital, which looked nothing like the more traditional positions in their portfolio. Early pioneers in the venture community such as Ben Webster of Helix, had demonstrated through their sustained commitment to venture capital the possibilities represented by backing entrepreneurs, and helping them to build new businesses. The large money managers began to set aside small sums to invest, typically for a large portion of equity ownership, in technology companies. However, with little experience in private equity, and often even less experience in the technology sector, calamity struck occasionally, wiping out much of the original capital placed. Soon this fledgling asset class became orphaned in Canada; only a few companies— like Ventures West in Vancouver—willing to tough it out and stay in the game of offering venture investment advice to the capital pools. During this era, there was no collective belief that technology would be crucial to building wealth. Canadian investors looked for returns on their investment in other sectors, such as natural resources. Technology became out of fashion. In this environment, it was extremely difficult to raise the funds required to support technological innovation. Most companies were forced to look elsewhere for equity funding, sometimes to the U.S., or a junior stock exchange.

Over the last 10 years, a new breed of venture capitalist has emerged in Canada. Funds like Jefferson Partners, McLean Watson Capital, Mosaic, and Sofinov have grown up with a more nimble and aggressive philosophy toward financing technology ventures. For many funds, having local domain expertise was a critical factor in attracting institutional money. Until this last decade, that kind of local knowledge was exceptionally limited. Firms like Brightspark, Aprilis, and Venture Coaches have added a level of domain expertise that was previously not within our national reach. The industry has evolved from a few players to one which has achieved substantial size, shown dramatic growth over the last five years, developed more and more capabilities, and has become a more attractive industry to top quality recruits.

Many entrepreneurs find that the right financing partner adds value over and above the distribution of money. Just like the problem mentioned above (which investors faced), the entrepreneurs themselves, until a decade ago, had very little ability to find "smart money." Today, however, Canadian venture capitalists have built up a string of successes (and many failures) that accumulate that priceless experience which entrepreneurs so desperately require at home. That cumulative experience is directed back to venture opportunities, increasing the odds of local success. The Canadian Pension Plan (CPP) has recently committed to grow it's capital allocation towards private equity, and is well on the way to becoming the largest private equity investor in the world. Along with other merging large players such as EdgeStone, Canada will have a substantial amount of capital available for backing entrepreneurs, and will lend substantial momentum to creating entrepreneurs in Canada.

Angel investors have truly moved out of the backroom and into the forefront of the financial landscape over the last decade. It has become an increasingly sophisticated art. Like venture capital firms, angels have money to invest in ideas and early-stage ventures. Angels, however, focus on the seed stage of a company, whereas venture capitalists tend to focus on companies that have already passed that critical stage. Angels are, in essence, a form of post-natal care. They tend to invest on their own behalf, not on behalf of others. As such they have a strong personal connection to the investment. As successful entrepreneurs return to Canada, they continue to strengthen the ranks of angel investors—to date, some of the more well-known angel groups or funds are Starting Startups in Ottawa, InvestAngels in Montreal, and Mike Volker's Vancouver Angel Forum. As this market matures, there is now a movement to create support networks for angels like Henry Vehovec's Mindfirst, based in Toronto.

The Canadian public markets have also enabled access to capital for early stage ventures. Not long ago, emerging technology companies would not meet eligibility criteria for the TSX, and would turn to the more speculative markets of the Montreal Stock Exchange, the Vancouver Stock Exchange, and the Alberta Stock Exchange. Subsequently, the public markets in Canada have changed fundamentally to improve their overall competitive stance, and have made substantial strides to accommodate the needs of entrepreneurs. We have seen a process of legitimization begin with the merging of the VSE and ASE into the CDNX, then the subsequent purchase of the CDNX by the TSX. At the time of writing, the public markets are integrating at a rapid pace, increasing access and making investor capital much more cost-effective. To date, the new CDNX, now the TSX

Venture Exchange, is really one of the few public venture capital exchanges of its kind in the world. Ongoing competitive pressures from other capital sources and markets will continue to drive change, ultimately benefiting the emerging companies. More importantly, over the last five years, the process has become increasingly more sophisticated and regulated, increasing public confidence, and thereby success, in the role of the public markets in venture capital in this country.

Finally, any look at the Canadian landscape of venture-backed investment would be incomplete without a mention of the labour-sponsored funds. These capital pools are another great example of the public-private partnership in this country. Firms like the Working Opportunities Fund in British Columbia and VenGrowth in Ontario allow for investment in venture deals while returning a tax credit to the investors in the fund, often offered on the public markets. Many of Canada's most successful startups, such as Algorithmics, Q9 Networks, fSona, and Inflazyme got their start with the help of labour-sponsored funds.

ACADEMIC BRIDGES—THE CRITICAL LINK BETWEEN UNIVERSITIES, CORPORATIONS AND BUSINESS BUILDERS

Like any family, the technology sector in Canada has its parental units. Since the founding of this country, the technology sector has been nurtured by two primary institutions—large corporations and our universities. A long-held view is that the rapidity of innovation increases when large companies slough off favourable research and development projects into separate, and more nimble, ventures. Bell Northern Research Labs had for many years funded and converted important research into commercial technologies. Scientists employed at BNR regularly left to establish and nurture businesses on their own. Of course, Nortel, Canada's well-known communications equipment manufacturer, has a long history of spawning new companies and encouraging intraprenuerialism. The engineers and scientists at Nortel would regularly leave to found new ventures—some in partnership with Nortel, and some not. One of the best-known offspring of the BNR and Nortel combination was founded in 1981. JDS FITEL, led by Jozef Straus and three colleagues from Nortel, focused on fibre optic communication and was eventually sold to Uniphase in 1999 for $7 billion.

Universities have long been the source of basic science and technology research in this country. They also had a similar effect on enabling new companies to develop and nurture. Some of the world's best science and engineering is pioneered and developed at Canadian post-secondary institutions, and, in turn, this often leads to world-class ventures. The ongoing pressures for universities to contain their costs, and to live within more restrictive economics, has accelerated their examination of innovation and business development to support their core mission. Many Canadian universities now have disciplined technology commercialization programs in place to accelerate the transfer of knowledge developed in their labs into commercial applications. Licensing agreements, consulting arrangements, and equity positions are increasingly the norm for universities to gain value from these efforts. This cycle will continue to change the way universities fund themselves over the next decade. One need only look to the University of Waterloo as a prime example of this trend.

Canada has tremendous wealth in its human capital, developing a taste for technology at an early age. For example, Canada was the first country to connect all of its schools and libraries to the Internet.[6] The country's talent continues to strengthen as Canada sends its students to post-secondary education at the highest rate in the world, at some of the world's finest institutions.[7] Despite our country's small size, a recent survey of electrical engineering programs in North America found that an incredible 18 of the top 40 programs were Canadian.[8] Additionally, Canada was ranked as number two in the world (behind the U.S.) in providing management education in business schools.[9] Tim Bray argues, "We have an absolutely world-class pool of engineering talent in this country. There's no place you can go to get better engineers than you can here." Don Mattrick, co-founder of Electronic Arts, goes on to say, "The one thing I've learned, and that we've learned as a

[6] Speech by Minister of Finance, to the Canadian Society of New York on Jan 17, 2001. http://www.fin.gc.ca/news01/01-007e.html
[7] The Government of Canada. Knowledge Matters: Canada's Innovation Strategy—www.hrdc-drhc.gc.ca/stratpol/sl-ca/doc/summary.shtml
[8] "The Gourman report." Dr. Jack Gourman—Undergraduate programs, 10th Edition 1998—as cited in http://www3.telus.net/info/ee-prog.htm
[9] Industry Canada. Arif Mahmud, Micro-Economic Analysis Directorate. Global Competitiveness Report. http://strategis.ic.gc.ca/scdt/invest/presentations/think/think!e.pdf

company, is that there are amazing people inside Canada. The quality of the people, their ability to excel, both technically and creatively, is one of the reasons why our company has succeeded the way it has."

The Canadian government has recognised the value of a strong educational infrastructure and has aggressively funded development in key areas such as research and development. For example, the government established the Canada Foundation for Innovation and has invested $3 billion since 1997 to fund research at Canadian universities and non-profit institutions.[10] This supplements 2,000 new research chairs that the government will establish over the next four years. These initiatives taken together, the universities ought to be able to continue their role as idea incubators, fueling the commercialization of innovation within this country. As Mike Lazaridis, co-founder of RIM, notes, "The federal government's Innovation Agenda is huge. The fact that they are willing to fund education, to fund the indirect costs of research at universities; they're mandating the growth of graduate programs. This hasn't happened in decades. This is a huge change. Assuming they can follow through with it and make good on their agenda, Canada will be a completely different place in ten years."

INFRASTRUCTURE BUILDERS—THE BACKBONE

Ask a Canadian about our technological success and you will likely hear two proud stories from years past, each of which ties to our powerful history in the infrastructure game. Canadians are quick to claim Alexander Graham Bell as one of our own, despite his move to the U.S. at the age of 23. They also note proudly that Signal Hill outside of St. John's, Newfoundland, was the site of the first transatlantic wireless signal in 1901. Like the railway, Canada's telecommunications networks have proved critical to our economic, social, and cultural development. Long before the term "information highway" was coined in 1993, the enabling effects of networks had been overtly recognised in this country. In the 1960s the Canadian government grasped the importance of the computer revolution and was convinced of the importance of "fostering the orderly development of national

[10]Canadian Minister of Industry's Keynote address to the Committee for Economic Development and eBusiness Roundtable on May 16, 2001. http://www.ic.gc.ca/cmb/Welcomeic.nsf/503cec39324f7372852564820068b211/85256a220056c2a485256a8f00561171!OpenDocument

telecommunications."[11] The Department of Communication's "Telecommission" and massive report "Instant World: A Report on Telecommunications in Canada" were strikingly prescient. It was clear from early on that without a proper technological backbone in this country, we could not scale our economy due to our inherent geographic constraints.

Regardless of where you stand on the regulation/deregulation spectrum, it seems that Canada's unique approach to "managed" deregulation has seen us establish a critical beachhead when it comes to telecommunications services. The often fractious struggle among incumbent telecom carriers; cable companies; the new entrants, whether in traditional services or in newer technologies; equipment manufacturers (e.g., Nortel and Mitel); and the policy-makers and regulators in government, has on the whole served Canadians very well. The men and women who led these companies through these exciting and difficult times have made important contributions to building our national infrastructure. People like Terry Matthews, who saw that the stranglehold Northern Telecom had on the telecom markets would be broken with the coming of interconnection of customer premise equipment and built his first company—Mitel—on that premise. People like Mike Kedar, of Call-Net, who challenged the giant telecommunications providers and kept getting back up when he was knocked down until he finally won the battle to compete in long distance markets in the early 1980s. People like Ted Rogers, who had a vision of an alternative "supercarrier" bringing together cable, long distance, and wireless services. People like the Simmonds family of Pickering who turned their family-owned business into a national wireless service provider—Clearnet—competing in the big leagues.

But equally important are the people who built the Trans Canada Telephone System. This made coast-to-coast long distance calling possible, bridging provincial regulatory barriers and steering the mammoth monopoly players into the new world. Also important were people who saw that significant investment in a national digital network running on optical fibre was vital, not optional. People like Jean Monty, then of BCE, and Carol Stephenson of Stentor, now of Lucent, who saw that the world was changing and that the large organizations they led had to innovate or die. People who realized that the business

[11] Laurence B. Mussio, *Telecom Nation: Telecommunications, Computers and Governments in Canada*, McGill-Queens, 2001.

could no longer rest on providing the conduit but required entering new and much riskier markets to provide value-added services.[12] People like the executives in the banks and the major users of services who understood the critical importance of telecom to their livelihood. This also applies to people in the governments and the CRTC who understood that communications infrastructure was critical to Canada's social and cultural development as well as its economic success. Innovation is not restricted to young startups. In fact, getting an 800-pound gorilla to move nimbly is quite a daunting task.

The delicate and shifting balance between free market forces and regulation has ensured that all Canadians have access to high-quality, affordable communications, leading the U.S. in many measures. Many innovators in small and large private sector companies, as well as in government, helped shape our national networks and support the vision of a "smart" country enabled by telecommunications.

Here is where our vision has taken us:

- Canada leads the world G8 in the penetration of basic telephone and cable services.
- Canadians are more connected than the population of the U.S.—60 percent of Canadians were online in 2001, compared to 52 percent of Americans.[13]
- Canada leads in government services online.[14]
- Canada is second in OECD countries in the rate of broadband use per inhabitant, at 4.5 per 100 inhabitants, almost double the U.S. (2.3 per 100).[15]

While some say that Canada has tended to lag a couple of years behind the U.S. in deregulation, others suggest that has actually been our greatest asset. Rather than being on the "bleeding edge," which can cut both ways, there has been more time taken to get ready. In addition, our market structure is less fragmented and has allowed the development, for example, of national wireless carriers, which do not really exist south of the border. No one would claim that the telecommunications markets are easy or stable. The infrastructure

---

[12] Andre Tremblay, "Competition in Telecommunications," *Economic and Technology Development Journal of Canada*, 1997.
[13] IDC, Internet Commerce Market Model, v. 8.1 (2002) as cited in Fast Forward 3.0 http://ebusinessroundtable.ca/
[14] accenture . eGovernment Leadership—Realizing the Vision. http://www.accenture.com/xdoc/en/industries/government/eGov_April2002_3.pdf
[15] Connecting Canadians, Key Statistics on ICT infrastructure, Use and Content, October 2001.

continues to have a voracious appetite for capital investment while yielding very narrow profit margins. There are also challenges ahead as a relatively small proportion of Canadian communities have access to interactive broadband services. Clearly for Canadian entrepreneurs to continue to succeed at home and abroad, we need to continue to focus our energy on continuing to develop our national infrastructure.

CANADA'S HOODS—THE TECH CLUSTERS

Similar to California's Silicon Valley, Boston's Route 128, and New York's Silicon Alley, Canada has a series of technology clusters that extend from coast to coast and account for roughly half the country's population. Each comprises a community of proven entrepreneurs, technology workers, universities, established technology companies, venture investment, and government support. These clusters can be seen as part of a national preoccupation with driving Canada's technological capabilities forward based on local areas of focus. We would argue that any one of these clusters, when matched up against its equivalents in the U.S., would meet or exceed even the toughest of expectations. While we recognise that there are many more clusters than those we have listed here, we have chosen to only highlight a few due to space limitations.

Ottawa—"Silicon Valley North"

Our nation's capital is Canada's most established technology community. The area has particular expertise in the fields of telecom, networking, and Internet infrastructure. It has become a favourite stomping ground for many U.S. venture capitalists; several firms are setting up joint partnerships in the National Capital Region to administer local investment. This is typically a key indicator of cluster maturity. For example, Silicon Valley's Newbury Ventures has teamed up with Ottawa-based Eagle One Ventures in just such an exercise. The more active venture capitalists in the region, in addition to Eagle One Ventures, are Celtic House, Venture Coaches, and Skypoint Capital. The two local universities of note are Carleton and the University of Ottawa. Some of the corporate residents of Silicon

Valley North include industry giants such as Nortel, JDS Uniphase, Accelio, and Corel, along with award-winning upstarts like Bitflash, Conexant, SS8 Networks, and Ceyba.

Toronto—"Silicon Alley North"

Toronto, the fifth-largest city in North America, is Canada's financial centre. The city is home to the University of Toronto, York University, Ryerson University and an unparalleled cadre of workers skilled in information technology, biomedicine, and biotechnology. Given the size of the market, the areas of specialisation in the region are quite broad and range across almost all genres that could bear any association with technology. Name the sector or industry, and it is guaranteed that there will be a homegrown world-class player in Toronto. Some of the more active venture capitalists in the region are Mosaic Venture Partners, JL Albright Partners, McLean Watson Capital, Brightspark, and Jefferson Partners. The city is head office to established firms such as Spar, Celestica, and Rogers, while housing a plethora of successful startups like Inquent, Cyberplex, Q9 Networks, Workbrain, and OpenCola.

Waterloo—"The Technology Triangle"

Out of the roots of the University of Waterloo (known as one of the finest technology schools in the world), the city has become a production line of successful companies. Waterloo has also become a magnet for public sector-funded research. The city is often grouped with Cambridge and Kitchener, which is where the "Technology Triangle" gets its name. This strong educational environment creates a flow of great engineers, employees, and role models for local companies. The result is a startup success rate that is extraordinary in relation to the size of the area. Even with the region's proximity to the financial community in Toronto, the strength of the Triangle has seen local venturists emerge led by seed fund Tech Capital Partners. Several impressive startups have been spawned from the Triangle—notably, Open Text, RIM, Pixstream, Sandvine, and Descartes.

Vancouver—"The Northern Tip of the Venture Silk Route"

There is rarely a person who has visited Vancouver who has not called it one of the most stunning cities in the world. That raw beauty is complemented by a rich technology sector. While the city sees activity in virtually all spheres, it is particularly strong in the fields of wireless, software, alternative energy, and life sciences. Vancouver is quickly becoming the northern tip of the venture silk route running from Silicon Valley through Portland and Seattle. The active local venture capitalists include Ventures West, Greenstone Partners, and Discovery Capital. The city is residence to two of Canada's leading research schools, the University of British Columbia and Simon Fraser University. In addition, Vancouver is home to senior technology firms such as CREO, Electronic Arts, Sierra Systems, PMC Sierra, Ballard Power, and Pivotal. It has also incubated such success stories as nCompass, Onvia.com, QLT, Inflazyme, Xenon Genetics, Convedia, and Antarctica Systems Inc.

Montreal—"Les Techabitents"

Almost everything about Montreal is unique. Not only is it known for its European flair and *joie de vivre*, it is also a top-rated cluster for biopharmaceutical research, aerospace products, and optical equipment. Its labour force is fed by talent from world-class McGill University and the Université de Montréal. The Quebec government has led all provincial governments in its efforts to support the technology community and encourage growth. Companies in Quebec benefit from programs that fund training and investment projects, and, in some cases, provide loan guarantees. There is a program that grants large tax credits to optics and photonics companies that move near Quebec City; another program offers income tax holidays for certain foreign R&D workers. Quebec has become a world leader in graphics and media. SoftImage developed a unique capability in 3D animation, and took Hollywood by storm with its technology capability, and spun off other important players such as Discreet Logic. Montreal is head office to one of the most active venture investors in the country—Sofinov—and houses other strong firms like Technocap and Bio Capital. BCE Emergis, Cognicase, and Bombardier call Montreal home, along with startups like Zero Knowledge, Galileo Genomics, Teraxion, Matrex and Hyperchip.

As Canadians learn the accelerating power of clusters, efforts are being made to more actively manage the creation of innovation hotbeds, where technology entrepreneurs can easily interact with angel and venture capital investors, accessible and demanding markets, and government support. The Canadian landscape is now networked and expanding.

CANADIAN LIVING

This brings us to another critical part of the Canadian advantage—Canada is a fantastic place to live. A company does not have to sacrifice its executive or employees' lifestyles to benefit from the financial upside of this country.

In 2001, the United Nations ranked Canada third on the Human Development Index, a comprehensive measure of quality of life factors, trailing only Norway and Australia.[16] Canada is, in fact, usually ranked first, holding the title for seven years in a row before slipping back two spots in 2001.[17] This reflects the solid healthcare and education systems, strong economy, low crime rates, and great cities of this nation. In addition, Vancouver recently tied as the best place to live in the world, according to the 2001 William M. Mercer Quality of Life Survey.[18]

This Canadian quality of life was another consistent theme raised in the profiles throughout this book. Many entrepreneurs built their businesses in Canada because they simply loved living in this country. Some, like Creo's Dan Gelbart and Brightspark's Tony Davis and Mark Skapinker, were immigrants to Canada. Each came with a view to utilizing the Canadian way of life to accelerate their ideas and hypergrow their businesses. Dan Gelbart, who moved from Israel, says he chose to live in Vancouver specifically for its serenity and its "isotherm of warm climates and no wars." Hammie Hill of Zero Knowledge adds, "for younger people, Montreal is a very, very cost-effective city, and it's also a big multicultural city. When we've brought guys out of California up here and some of them have actually moved up with their families, they're amazed they can walk down the street with their wives

[16] United Nations Development Programme. Human Development Report. http://www.undp.org/hdr2001/hdi.pdf
[17] The Government of Canada. Canada's Winning Secrets. http://www.gov.mb.ca/itm/cc/location/investincanada_feb21.pdf
[18] Vancouver was tied with Zurich for the top spot in the William M. Mercer Quality of Life Survey (2001). As cited by the City of Vancouver—http://www.city.vancouver.bc.ca/ctyclerk/newsreleases2001/NRvankudosmercer.htm

at ten or eleven o'clock at night and not even have a second thought about their safety."

Live in this country for six months and you may never want to go home.

THE CANADIAN IMPERATIVE

Over the last decade, Canada has undergone a transformation of incredible proportions. Taking heed of the warnings of economic downfall, the country, both public and private sectors, turned the boat around. Through a combination of recognising our natural assets and changing the major obstacles in the way of growth, Canada put the country at the forefront of the technology revolution. This country now has a very hospitable and internationally competitive environment for business.

Foreign investors would be well served to look north of the border to take advantage of the unique combination of easy access to the NAFTA market, extremely low costs of doing business, and an unparalleled quality of life. A dollar of venture capital will simply buy more in Canada than it does in the United States. Canadian companies had more reasonable valuations and were better at keeping their burn rates down even during the height of the Internet bubble. Foreign investors maintain their strong interest in Canada, notwithstanding the equity market pullback, and have increased their commitment to Canadian companies in a sustained way over the last two years.

Foreign investors also recognised the value add available in Canada. Foreign investors have continued to commit to Canadian companies. This statistic reflects the confidence that existing investors have in the Canadian market. Investors also recognise that Canada provides an easy method of increasing their portfolio diversity. Canada delivers a foreign and geographical solidity in an area of the world that the investors can still understand and monitor. Canada and the U.S. are the world's largest trading partners; over $1 billion in goods cross the border every day.[19] The two countries enjoy the low trade barriers of NAFTA and Canadian cities are well positioned to access the U.S. market. Toronto is within a two-hour flight of more than half the U.S. population.[20]

[19] Innovation Canada. Ian Burchett, Consul, Investment and Corporate Relations and Robert Bult, Investment Officer; Canadian Consulate General, New York. www.locationcanada.com/article_growth.asp. See also: The Government of Canada. Canada's Winning Secrets. http://www.gov.mb.ca/itm/cc/location/investincanada_feb21.pdf
[20] Industry Canada. Arif Mahmud, Micro-Economic Analysis Directorate. Global Competitiveness Report. http://strategis.ic.gc.ca/scdt/invest/presentations/think/think!e.pdf

Canadian companies enjoy a striking cost advantage. In 2002, KPMG completed a comprehensive study of industrial nations in Europe and North America, along with Japan. They concluded that Canada boasts the lowest costs of doing business of any country in the study, with a 14.5 percent cost advantage over the United States.[21] This cost advantage is even larger in some key technology industries. Canada has a 22.3 percent cost advantage over the U.S. in software, and an enormous 30.6 percent cost advantage in research and development.[22] According to Tim Bray, the costs of doing business in Canada are just immensely lower. Everything, "the computer bandwidth, the cost of office space, the cost of people, you name it, it's cheaper here."

The magnitude of Canada's cost advantage can best be understood by looking at the individual cities in the KPMG survey. Most people would not be surprised that Toronto, Montreal, and Vancouver are less expensive than New York, San Francisco, and Boston. However, most Americans would be shocked to find that these world-class Canadian cities are significantly less expensive than Lexington, Kentucky; Lewiston, Maine; Cedar Rapids, Iowa; or Sioux Falls, South Dakota.

THE FIRST PERIOD OF A GAME DESTINED FOR OVERTIME

What we hope this book will demonstrate is that Canada is truly among the upper echelons of the global technology market. Our inherent humility, in relation to the U.S. in almost all commercial activity, is not warranted here. The last 10 years have seen a dramatic turnaround, a "harnessing" of sorts, whereby the country turned a long-standing love affair with technology into actionable leadership. However, we are also confident that the future will be even brighter than the present. This book is only the beginning of the story. We are in the first period of a game destined for overtime.

There has been a general increase in the appetite for, and understanding of, the importance of technology and innovation in Canada. Canadian businesses and government have become more comfortable with technology, as evidenced by the increasing

[21] Note: The KPMG study used an exchange rate of $1 U.S. = $1.546 Can. Since the time of the report, the exchange rate has shifted and the Canadian advantage is now even larger. KPMG Competitive Alternatives 2002. http://www.competitivealternatives.com
[22] KPMG Competitive Alternatives 2002. http://www.competitivealternatives.com

number of businesses with an Internet presence and the aggressive Government Online (GOL) initiative that includes a $600 million dollar funding commitment over four years. Beyond this general movement, there are also several specific trends that we believe will continue to drive accelerated change and success in our Innovation Nation. There is much reason for optimism.

Governments Must Stay the Course: Substantial change for the better has occurred within the government sphere in creating a welcoming, enabling environment for technology and innovation. We must ensure that the ice is maintained and that we have a deep bench of capable players. Our taxes must continue to be competitive with rates in the U.S. and our talent pool must continue to be developed. The Federal government has continued to make substantial strides in making Canada more "friendly" to foreign capital to support the development of our entrepreneurs. This represents a substantial change in philosophy and commitment to improving the economic health of Canada. The good news is, however, that the Canadian government now clearly understands the equation that technology + innovation = productivity. Based on this belief, Ottawa has named innovation as a national goal and has officially launched an ambitious Innovation Agenda that will roll out in the fall of this year. We are confident that our leaders will continue to select the policies that will drive technology forward at an ever-accelerating pace.

Governments at the provincial and municipal levels are also initiating changes to encourage technology innovation and business building. Many of the clusters in Canada are the subjects of intense examination by local governments to understand how better to support their development.

Becoming Wealthy in Canada: While early-stage capital is increasingly available, even more is required to fuel the growth of our next wave of startups. The good news is that the Canadian venture capital industry has matured and the increased sophistication of investors has led to better partnerships with the nation's entrepreneurs. Canada now welcomes foreign investors with open arms. This greater openness will lead to an increased availability of capital and advisory expertise for our nation's startups and helps keep them growing and creating their wealth at home.

Post-Secondary Education Rediscovers Business: Universities and colleges have played a key role in providing talent, but increasingly they will also be players in commercializing new technologies and other innovations. As some of the best educational facilities in the world, they continue to accelerate innovation and are experimenting with private partnerships like the University of Toronto's Medical and Related Sciences Discovery District (MARS). This research park is scheduled to be built in downtown Toronto and has been funded by private and public sources to accelerate commercial applications for medical and university research.[23] A similar research centre called UW Research and Technology Park is under consideration near the University of Waterloo.[24] These partnerships with the private sector come at a time when businesses have begun to realize the value of universities and their intellectual capital. We are likely to see private companies working hard to advance these relationships in the near future. Meanwhile, in the background, our post-secondary institutions are churning out innovative research and entrepreneurs at an unprecedented pace in our history.

Clusters Deepen and Broaden: The communities of interest, which have developed in Canada over the last decade, are now hitting the accelerator. Networks are becoming more intertwined and critical mass has been reached. The virtuous cycle that drives these technology clusters will continue to hasten and all aspects of our technology hubs will be strengthened. The virtuous cycle that is often used to explain the dramatic success of Silicon Valley and Boston's tech sector will also continue to deepen the technology hotbeds of this nation. In this book we didn't even have space to mention the new clusters emerging in cities like Calgary and Halifax.

Canadians Come Home: Many Canadians who have battled competition in the Silicon Valleys of the world have begun to, or will soon, return home. As we have seen, some of these serial entrepreneurs bring capital, but more importantly, they bring deep experience, management expertise, and a history of accomplishment. These walking success stories will help to inspire and develop the next generation of Canadian entrepreneurs. The troops are coming home and bringing their networks and contacts back to the benefit of the nation.

[23] University of Toronto Magazine. http://www.magazine.utoronto.ca/01spring/campusnews.htm
[24] http://www.region.waterloo.on.ca/web/region.nsf/c56e308f49bfeb7885256abc0071ec9a/8648d4d8bee1e31b85256b1f004b08a6!OpenDocument

GOING FORWARD

The points raised in this introduction could be the subject of an entirely separate book, and we hope that they will prompt discussion and debate. The true focus of this project, however, is on the people—the men and women who have demonstrated that Canada has made it possible for them to be inspired, live, create, and grow their enterprises. We know that we have only scratched the surface in terms of the people who should rightfully be listed here. We truly wish we could have had room for everyone; however, we think we have chosen a great first testament to the leaders of our technology sector. We are confident that there are enough stories out there to work on Innovation Nation II, III, and IV. This is really just the beginning—the genesis.

Going forward, now that the secret is out, it is our hope that collectively, in the Canadian spirit, we continue to steer the boat in the direction we have so boldly headed. The perspective of this book has to be longer term than the most recent 12 months. At any time over the last several years, a short-term perspective would prove absolutely wrong as a predictor. Capital markets may roil, and the technology stocks may be on a roller coaster from one day to the next, but it is in our national interest to ensure that we do not waver. Technology is the future of this country and innovation is the path that will take us there.

So, grab a Molson's, a Labatt's, or a Moosehead. Put it on the table next to your tourtière, your fish and brewis, some bannock and a beavertail (killaloe sunrise, no doubt) and get ready to see how 30 Canadians quietly became architects of the technology economy. We truly hope you enjoy reading these stories as much as we did interviewing, learning and writing about them.

Leonard, Wendy, Ken, Matt, Catherine, and Denise

ENTREPRENEUR IN THE EXTREME

INNOVATOR_ GLENN BALLMAN, SERIAL ENTREPRENEUR, COACH, FINANCIER INNOVATION_ RAISED CLOSE TO HALF A BILLION DOLLARS TO BUILD THE FIRST AND LARGEST ONLINE MARKET FOR SMALL BUSINESSES IN THE WORLD

He may have grown up in small-town Saskatchewan, but thinking big has always been a mantra for Glenn Ballman. At the age of 17 he committed to the idea of building a large multinational before he reached age 29—an admittedly odd goal for a teenager, but Ballman accomplished this enormous feat a year ahead of schedule, with the success of his company, Onvia. An online retailer/portal for small business, Onvia's was one of the largest Internet IPOs of 2000. On the day of Onvia's IPO, Ballman's paper wealth was valued at more than $800 million (U.S.).

For Ballman, size is everything, and he'll tell any entrepreneur who'll listen to dream big. "If you want to start something, it takes just as much effort to run a corner store as it does to run a multi-billion-dollar multinational. You're going to work 18 hours a day anyway, so choose your battle. Make it big. Make sure it's going to be a huge market and build your plan to be the number-one company in that market," he says. Colleagues have described Ballman's drive and determination as "superhuman" in an industry defined by drive and determination.

After he graduated from the University of Western Ontario's business school, Ballman embarked on his entrepreneurial journey with Megadepot, a small startup he founded from his home in Vancouver in 1996. Megadepot.com was a website where entrepreneurs could meet in cyberspace to buy and sell products and services and to access other resources, such as business information and productivity tools. In 1998 Ballman moved his headquarters to Seattle, Washington, in search of venture capital and a larger market. Working punishingly long hours and making many sacrifices, he grew his company at an impressive rate, quadrupling his staff from 15 to 60 people in 90 days and expanding its office space from 1,200 square feet to 20,000. Eventually this company changed it's name to Onvia and boasted more than a million users and thousands of suppliers registered on its Website. It also formed strategic business alliances with major companies, such as VISA and AOL, and with dozens of small business associations and chambers of commerce throughout North America.

While the company's growth was astronomical, it was by no means easy. The first time Ballman flew to New York to seek funding for his business he was unceremoniously sent packing without so much as a call back. He didn't let that traumatic experience deter him, though he admits that wasn't the last time he heard that discouraging message. "When I first moved to the U.S., I was told that I had no business raising money. In their opinion, I just didn't have the straight goods, and I should step aside as CEO. I heard this from a few VC firms—and in some ways they were right, since I had never done it before. But they were only partially right. You can't measure somebody's ability and willingness to learn through a couple of meetings. I ended up raising $300 million just for Onvia, but it's a learned skill like everything else." In total Ballman took the company through five rounds of private financing before taking it public a year later.

Where you raise the money is also key, continues Ballman. "Most people go to the banks first, but if you track entrepreneurs who went from 'zero to hero,' it's about where they raised their money, and it's very seldom from the first ten places that you approach. It's actually the eleventh try, the twelfth or the twentieth try. And they finally figure out, aha, this is the niche group of people that understand my business, can add value to my company, and are interested enough to put money in."

"IF YOU WANT TO START SOMETHING, IT TAKES JUST AS MUCH EFFORT TO RUN A CORNER STORE AS IT DOES TO RUN A MULTI-BILLION-DOLLAR MULTINATIONAL. YOU'RE GOING TO WORK 18 HOURS A DAY ANYWAY, SO CHOOSE YOUR BATTLE. MAKE IT BIG. MAKE SURE IT'S GOING TO BE A HUGE MARKET AND BUILD YOUR PLAN TO BE THE NUMBER-ONE COMPANY IN THAT MARKET."

Eventually, Ballman took his small-town company to extraordinary heights, and for a brief, shining moment in history Onvia was worth $6 billion (U.S.). He says the secret of achieving this kind of scope is simply surrounding yourself with people who are smarter than you are. Ballman suggests this lesson may be particularly hard for most entrepreneurs to learn, especially those who are ego-driven. As CEO, Ballman was proud of the fact that most of his team was smarter than he was. Onvia's own dream team included managers, advisors, and employees recruited specifically for being the best in the world at what they did: "the kind of people with professional expertise, who can scale a small company into a billion-dollar business." As leader, Ballman insisted that everyone who touched his business be "best of breed," including Onvia's board of directors, outside partners, and affiliates, so that ultimately the company came to be "built through the relationships it had with world-class professionals, and would by nature eventually become world-class itself."

> **"THINK BIGGER THAN THE GUY BESIDE YOU, AND BE PREPARED TO SACRIFICE EVERYTHING!"**

For example, when it came time to choose which one of three investment banks to go with, Ballman picked the one that was the best in the world in technology at the time. The same approach was used when signing advertising and PR firms, and even suppliers. Ballman says this approach is "a very different way to build a company; most entrepreneurs build by convenience"—that is, picking the supplier down the road or choosing a partner via personal networks rather than through careful research.

Ballman's "best of breed" philosophy also meant that he was not just building a company but creating a lifestyle for himself and the people in his company. He says that was his idea of true success. "If you can create your own lifestyle and it's centred around principle-based economics, such as 'serve the customer, fill a niche, make a profit,' you have the freedom to do what you want to do as well as to compete and be strong." This winning approach may have been what secured Onvia.com's position as the number-one startup of 2000 in *Profit* magazine and got the Canadian branch voted "the coolest place to work in Vancouver" by *Business in Vancouver* magazine.

For Ballman, it was never particularly about the money. Even in the heady days of Onvia, his lifestyle remained relatively austere. He shared a West Seattle apartment with three roommates. Not that he did not buy toys, including his own airplane, which he refers to as "a Winnebago with wings." But Ballman has been nonchalant about the rise and fall of his personal fortune.

When he left Onvia in 2001 he had numerous startup projects underway. "My passion lies in building companies from scratch," he says. After having competed and won in the global arena, Ballman is now sharing his ample wisdom with other budding entrepreneurs. Last year, he even held bootcamp-style summer sessions for friends and colleagues at his cottage in rural British Columbia. Referred to as "the Shawinigan Lake Club," this informal setting helped seven separate companies strategize for entrepreneurial success. The first lesson focused on the all-important step-by-step process of fundraising and on the instruction to treat potential investors like customers. He says, "Raising money is a sale. You have a company and you're selling the notion that this company is going to be successful, and so there's a sales process, and there's a closing process. You begin by learning the vernacular or vocabulary and then you must get familiar with the mechanics of a prospectus before you can move forward." It's no wonder wannabe entrepreneurs are flocking to Ballman for guidance, since Onvia's own prospectus was the most-downloaded in history at the time the company went to market.

During the "work hard and play hard" cottage sessions, Ballman draws repeatedly from his experiences with Onvia, where he handpicked employees who were not only top calibre, but also shared his extreme mindset and entrepreneurial agility. "It was a company full of triathletes, long distance runners, mountain bikers, and hockey players. We'd sooner burn out and collapse at the end of the day than just be OK players." Approaching business prepared to achieve a personal best of Olympic proportions has always been Ballman's style, and he sums it up with one last piece of advice for the entrepreneur: "Think bigger than the guy beside you, and be prepared to sacrifice everything!"

Jim Balsillie Michael Lazaridis_

YOU CAN'T LEARN SURFING FROM A TEXTBOOK

INNOVATOR_ JIM BALSILLIE AND MICHAEL LAZARIDIS, SERIAL ENTREPRENEURS, ACADEMIC BRIDGES INNOVATION_ DEVELOPMENT OF THE BLACKBERRY, WHICH HELPED ESTABLISH THE CONSUMER MARKET FOR WIRELESS DEVICES

Research in Motion Ltd. was founded in 1984 and became a "15-year overnight success" when it launched Blackberry in 1999, which soon became the world's leading handheld wireless e-mail device. Blackberry exploited the intersection between two burgeoning technologies —wireless and the Internet. As people got more and more hooked on e-mail and accustomed to the mobility of their cellphones, Blackberry helped establish a huge consumer market for wireless data with over 300,000 devices deployed. It has also become one of the fastest-growing high-tech companies in the country: revenues exceeded $500 million and profits more than doubled in 2001. In that year, the company's staff also doubled. A recently announced deal with Motorola Inc. and Nextel Communications Inc. to collaborate on a device offering both e-mail and voice has huge potential to expand their access to markets.

The success of the company is the result of strong technology, persistence, and a unique combination of rigour and intuition. Rigour is necessary, but insufficient on its own. You need the intuition to lead you to things that nobody else has considered. As co-CEO Jim Balsillie points out, "If it was so obvious, then somebody else would do it. Once something becomes conventional, by definition, it's no longer a good idea. It's only through something unconventional that you have a chance to make something."

Mike Lazaridis started the business, in his words, to "build a factory next to the gold mine." The gold mine was the University of Waterloo, a wealth of creative and technical talent. Initially RIM's focus was traditional wireless data. The challenge was not, fundamentally, about technology; it was about its applications. "We had to find an application that took advantage of wireless data, that was compelling. We realised that there had only been four successful wireless applications in history: paging; broadcast television and radio; cellular phones; and point-to-point microwave links and satellites. That's it. Everything else is niche, like radio-controlled cars and garage door openers. We built a matrix and we tried to find something that would utilise the value in one of those quadrants. We chose paging because people were paying enormous amounts of money to be notified that an operator had a message for them. We realised if we could make e-mail available 24/7 then we'd have married the value of paging with the utility of e-mail. That's what Blackberry is all about."

"It was a brand new idea; nobody had experienced it before, so it was very hard for them to perceive the value of reading e-mail on a tiny little device. We found in focus groups that if you try and describe the problem, using the best techniques you can, and the solution, nobody gets it. Once we got the whole thing working and put the devices in their hands, they didn't want to give them back. Once they had actually tried it and sent an e-mail to a friend and immediately got a response—they're hooked!" says Lazaridis. When it comes right down to it, Lazaridis maintains technological innovation "is about solving problems. Infact, just about all business is about solving problems."

It took a number of runs before the company's big breakthrough with Blackberry. "Wireless data has been around for a long time, but it only really took off in the last five years. Being in the business as long as RIM has made a huge difference in our ability to learn what works and what doesn't. A lot of the technical success behind Blackberry has been a fundamental, profound understanding of the limitations of the wireless data medium, the devices that are possible with the existing technology, anticipating what will become available and solving a real problem. Because, quite frankly, wireless data sounds exciting, but if it doesn't solve a real problem, why would you pay for it? So for us it was a discovery that took many years." For a while, the company abandoned wireless data entirely. "It wasn't until we told ourselves there was no 'wireless data' business, that in fact it's all about solutions, that we had the courage and the motivation to put together our complete Blackberry solution."

It is critically important to keep focused on the applications and on the business case rather than getting "romantically involved with the technology. The big challenge for us now is to stay on that path without being caught up in the old game, which is, 'Well, if we got a little bit more bandwidth, we could do video and all this other stuff.' We've been able to keep ourselves from running on the rocks that many companies out there have crashed into, by just holding meetings in our offices where we have calculators and paper and whiteboards and we actually calculate the feasibility. We don't allow ourselves to be mesmerised by how exciting it would be to have video conferencing on your cell phone."

Lazaridis confides, "One of the secrets actually is that if you're in there long enough, you end up successful. Statistics show that you will end up being in the right place at the right time, at least once. Longevity is important, it's not going to happen overnight. The cases where it does are statistically rare." Balsillie adds, "There is also absolutely no substitute for determination and persistence and hard work. It's true that it has to be an obsession. And if you love it, and you're good at it, that's good, because you're going to be living it for a long time; so you better like it. There's no substitute, in my mind, for obsessive hard work and determination. The dot-com phenomenon may have done everyone a disservice, because it brought too much success too soon; there's that sense of entitlement. Your whole expectation metric changes."

First and foremost, Lazaridis insists, "We do the math. We believe in what the math's telling us. In business, or in technology, you have to believe what the math's telling you." Balsillie, with a coveted Harvard MBA as well as a Chartered Accountant designation, agrees. "You have to be smart; you have to be rigorous. Nothing is a substitute for rigour. If you do your analysis, you do your homework, you understand the facts and the physics and the realities and the economics and the pressure points of your circumstance, you may not get the right answer, but you'll be very close." At the same time, he emphasizes, "intuition is not something they can teach. I don't think you learn how entrepreneurs manage; I think you learn how to think differently. Can I learn to think? To see the patterns from a different perspective, different dimension? And so you see things, very differently, very unconventionally."

> **A LOT OF THE TECHNICAL SUCCESS BEHIND BLACKBERRY HAS BEEN A FUNDAMENTAL, PROFOUND UNDERSTANDING OF THE LIMITATIONS OF THE WIRELESS DATA MEDIUM, THE DEVICES THAT ARE POSSIBLE WITH THE EXISTING TECHNOLOGY, ANTICIPATING WHAT WILL BECOME AVAILABLE AND SOLVING A REAL PROBLEM.**

"Business strategy must be developed in the context of a dynamic of the momentum of a wave or a current, rather than this sort of static set of things. You learn that by riding it and feeling it. You don't learn surfing from a textbook. You learn surfing by being on a board and being scared to death and being knocked on your head a bunch of times. Then one of the times you actually didn't get knocked on your head and it added up. For some people it adds up sooner, for some people it adds up later, and for some people it never adds up. But if you don't try, you're assured that it never will add up."

Execution is about sensing and it's about trust, it's about teams; it's about communication. And both Lazaridis and Balsillie insist that RIM's success is a result of access to talent. "The best thing that I've ever done, and continued to do, is hire the people around me. You have to handpick them; they're absolutely essential to your success. You have to have a great team and you have to keep them focused. The most important decisions you will make are who you choose for your partners, and who you choose for your employees. Up until we were about 500 people, I was personally involved in every hire that we made," says Lazaridis.

Balsillie adds, "Operating in Canada creates a particular context. It's got its own unique attributes, and so the key is to create a strategy around your own set of unique attributes. It's kind of like saying, 'Are big strong heavy legs an asset?' If you're a figure skater and not a speed skater, you've not picked the right sport and you're not doing the right training. You need to be able to leverage Canada's unique strengths. The big strong legs of Canada are multifaceted. We have a particularly well-educated set of workers. That is why RIM is adjacent to the University of Waterloo. Access to skilled people is a big advantage. The proximity and openness to the U.S. markets is key. Waterloo is closer to Atlanta than New York is. We're as central as anything in the U.S., so we have access to the whole U.S. market."

Lazaridis says, "I think that being in Canada is challenging but I think that it's a great place to be, if you can be successful, because you really wouldn't want to raise a family anywhere else. There is more opportunity in Canada than anywhere else. I think it's easier to start a company in Canada than it is in the U.S. and be successful."

And neither Lazaridis or Balsillie underestimates the importance of government policies. Lazaridis sees support for developing "human capital" as key. "The federal government's Innovation Agenda is huge—the fact that it is willing to fund education, to fund the indirect costs of research at universities; they're mandating the growth of graduate programs. This hasn't happened in decades. This is a huge change. Assuming they can follow through with it and make good with their agenda, Canada will be a completely different place in 10 years."

THE SEER

INNOVATOR_ TIM BRAY, SERIAL ENTREPRENEUR, INFRASTRUCTURE BUILDER INNOVATION_ CO-AUTHOR OF XML, AND THE ONLY CANADIAN ON THE PRESTIGIOUS TECHNICAL ARCHITECTURE GROUP OF THE WORLD WIDE WEB CONSORTIUM

Extensible Markup Language (XML) has revolutionized the way information is stored, searched, and retrieved on the World Wide Web. XML has become the universal format for structured information, an efficient way to ship objects from one point to another across different network platforms. And Canadian Tim Bray was one of the developers of the standard from 1996 to 1998. He is known worldwide as "the XML guy" and he is certainly proud of that. "Despite the billions of dollars and lifetimes of effort that have been poured into the Web, it looks and works about the same as in 1995. This can't last and XML has a big role to play in several generational changes that are coming" he says.[1] But Bray is also clear-sighted about the limits to what the standard does and does not do. He is concerned about "the hype problem," about those who tout XML as a cure-all; those who "are practically saying that XML will bring peace to Bosnia and cure cancer." Bray insists that when it comes right down to it, the Web still could work better.

[1] Tim Bray, "XML: Three Letters that every CIO should know," *CIO*, July 1, 2002.

Bray is widely regarded as an Internet pioneer and has enormous credibility in the development world. He was nominated by Tim Berners-Lee, inventor of the World Wide Web and director of the Web Consortium, to the Consortium's Technical Advisory Group, which oversees the architecture for the whole Web. Bray routinely globe trots to share his vision of the potential of the Internet and plays an active role in shaping the future. Like many pioneers, he wants to go where no one has gone before. But he's also helped map the terrain for those who follow in his footsteps. He has made his mark on search engines, on software standards, and now, at Antarctica Systems Inc., on visualization tools.

After a degree in computer science at the University of Guelph, he "stumbled" into computing jobs with Digital Equipment Corporation (DEC) and, subsequently, GTE. In 1987 he was charged with developing an indexing technology for the New Oxford English Dictionary Project at the University of Waterloo, in order to put the contents of the *Oxford English Dictionary* online, and make it searchable. It was a job he describes as "about as much fun as you can have and get paid for."

Building on his developing "way with words," in 1989 he co-founded Open Text Corporation (NASDAQ:OTEX), where he developed high-performance text retrieval software. In 1994 the company introduced what would become one of the first commercial Web search engines. "Our stuff was really Internet-oriented before the Internet was much of a thing. In 1994 the Web came along and all of a sudden our technology was positioned exactly where the world was going."[2]

He left the search engine development company in 1996 to set up his own consulting business. At a friend's request, he agreed to sit on the international Internet standards committee that developed XML. When he was co-editing the XML specification he had no idea that it would succeed beyond anyone's greatest expectations.

But after awhile, Bray tired of the standards game. He says, "Writing computer software is harder than building standards." So he shifted his focus to developing software, because he felt there was still a lot of room for improvement, particularly in terms of interfaces. Like any explorer, he firmly believes that there is a need for better maps, guidebooks, and pictures to allow users to navigate cyberspace.

[2] Susanne Hasulo, "Up Close: Tim Bray Mapping the Web," *Information Highways*, April 2, 2002.

He sees the next step in the evolution of the Internet as the user interface. He is now the CTO of Vancouver-based Antarctica Systems Inc., a company that is developing information visualization technology for the Internet. His goal is to create better user interfaces for data, using visual mapping technologies, which he sees as providing a better metaphor for the needs of networked information users. "A typical computer desktop is all based on windows and icons and wastebaskets and folders and so on," he says. "But as soon as you get off of your desktop out into the network, you're back into the mode of typing things into a box and hitting 'Enter' and looking at lists. And it just doesn't seem to make sense that personal data is visual but shared data is textual. Why should it be that way?"

LIKE MANY PIONEERS, HE WANTS TO GO WHERE NO ONE HAS GONE BEFORE. BUT HE'S ALSO HELPED MAP THE TERRAIN FOR THOSE WHO FOLLOW IN HIS FOOTSTEPS. HE HAS MADE HIS MARK ON SEARCH ENGINES, ON SOFTWARE STANDARDS, AND NOW, AT ANTARCTICA, ON VISUALIZATION TOOLS.

In some ways it makes perfect sense to go graphical in order to navigate cyberspace. "I always felt the Net has something of the feel of an urban landscape." Antarctica's software can be used to map any data set that has "some sort of an organizing scheme that allows us to lay it out on a map in an interesting way." As Bray says, that can be almost anything, which means there is a huge potential marketplace for Antarctica's services. For example, the company has developed a system that maps venture capital deals done in Canada. Rather than looking at a text-based list of these deals, the VCDeal Map plots venture capital activity graphically, making it easy to see which sectors are most active (based on the size of icon representing each sector) and to drill down to see details on individual deals.

With a new generation of technology users emerging who were weaned on navigating game space in three dimensions, the notion that users can better navigate collections of documents if data is shown spatially may be an idea whose time has come. GrowthWorks, BDC, Royal Bank Capital Partners, and other investors put more than $3 million into the project in the startup phase. But the challenges of selling the concept involve a major paradigm shift in how people think about the Net and information.

Although Antarctica's initial plan was to focus on developing a base of enterprise clients, the company's clients now include medical researchers and research libraries, groups that were not initially part of a targeted marketing plan. Bray comments, "Like every small business, we are continually re-evaluating ourselves and re-evaluating the market and finding the best way forward." Like many successful serial entrepreneurs, Bray believes that this ability to respond to the marketplace is a key to success. Once an explorer understands the terrain, he or she often discovers that some parts of it are more hospitable than others. "If you look at most successful technology businesses, you will find they are not doing today what they thought they were going to be doing when they started. It takes some intelligence and initiative to build useful technology. It takes divine gifts of foresight to predict where the technology's value proposition is going to play well in the marketplace. And in fact, I don't know anybody who can predict that reliably, so as far as I know, the only way forward is to get out there and talk to the marketplace and show your wares, and learn, from people you talk to, where the stuff works." Just like exploring—the best way to understand an environment is to experience it.

But not everyone is an explorer. In most businesses, someone already knows the terrain. Bray thinks that it's essential for anyone who is starting a new business to draw on that experience. "It's immensely important to have somebody on your team who's done it before. There are a lot of lessons, from basic practical things, like how do you structure your shares, to much more vital things like how do you get the attention of the marketplace." An experienced entrepreneur can provide guidance. In 2002, Bray was named a member of *Upside* magazine's "Elite 100," a select group of business leaders deemed the most powerful and influential in North America. He was recognised for his "hands-on approach in leading his company, and for creating technology that affects the way people access information."

Like many successful Canadian entrepreneurs, Bray could easily pack up his business and move it to the U.S. However, he's not leaving Vancouver. "The largest market in the world for technology is the United States, and Canada has a pretty good platform to sell technology into the United States. In Canada, the costs of doing business are just immensely lower. Everything, the network bandwidth, the cost of office space, the cost of people, you name it, it's cheaper here.

"We have an absolutely world-class pool of engineering talent in this country. There's no place you can go to get better engineers than you can here. Vancouver is a really nice place to live, and that gives us an advantage in terms of the talent pool available. There are a lot of people who are eager to come and live here." And although some Canadians gripe about the country's tax structure, Bray notes, "A lot of people don't know that the capital gains tax laws in Canada are actually substantially more entrepreneur-friendly than the basic income tax law is. I would argue that in some respects, for the entrepreneur you have a good chance of taking home a higher proportion of your capital gains than Americans would."

Canada is where Bray wants to be. He says, "The sort of work that I do is interesting and I love it, but it's still only part of my life. When I'm not at work, I'm just being a person, and Canada offers really excellent quality of life. And I think that living in a place where you enjoy your basic day-to-day life probably makes you not only a better person, but a better businessman too. So the quality-of-life tradeoffs are pretty good." Canada is lucky that Bray does his exploration in virtual space, while remaining firmly grounded in Vancouver. Bray's contributions to improved navigation and mapping of virtual spaces provide a solid foundation from which future explorers can begin their quests.

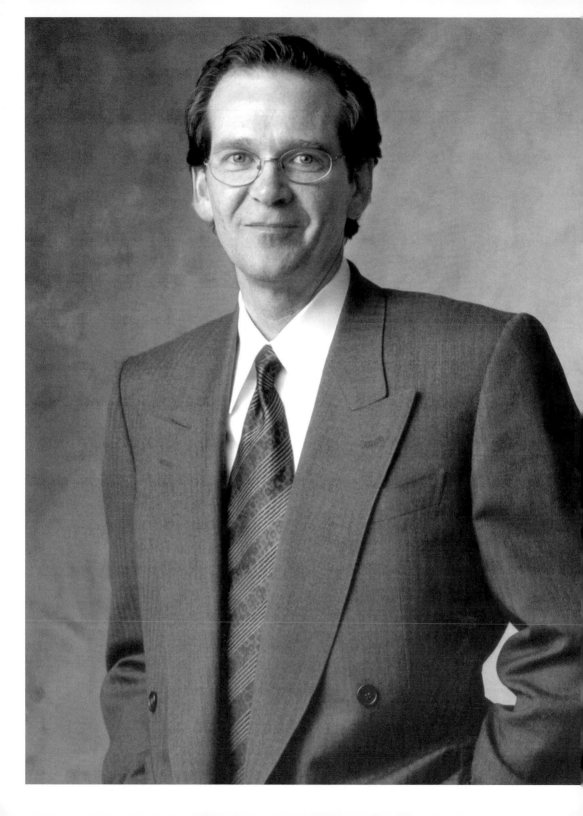

ROB BURGESS "The Internet is still a very nascent place. Most of the things that have been invested in, from an Internet perspective, don't work, which makes sense because people have only really been at it for three or four years. You can't make complex systems in that length of time. So there's really a great opportunity there."

THE ART OF THE TURNAROUND

INNOVATOR_ ROB BURGESS, SERIAL ENTREPRENEUR INNOVATION_ RESTRUCTURED AN AILING SOFTWARE FIRM INTO THE MOST INFLUENTIAL PROVIDER OF SOFTWARE TOOLS TO THE WEB DESIGN AND DEVELOPMENT COMMUNITY IN THE WORLD

Anyone who wants to run a profitable technology company can learn a lot from Rob Burgess. Burgess knows that keeping your focus on the game is key to a winning strategy. Presently chair and chief executive officer of San Francisco-based Macromedia, Burgess gained his reputation as a corporate turnaround artist at Toronto's Alias Research, a 3-D animation software company. In 1991, when Burgess joined Alias as president, it was close to bankruptcy. Less than five years later, Alias had become, in Burgess's words, "the largest and most profitable 3-D software company in history." The company was acquired by Silicon Graphics for $460 million (U.S.) and merged with Wavefront to create Alias|Wavefront.

How did Burgess engineer this turnaround? By focusing. Burgess's strength as a manager is in helping companies understand what they are good at, and then focusing all their energies on doing it better. At Alias Research, this meant developing best-in-class animation

> **BURGESS MAKES IT LOOK EASY FOR A COMPANY TO DEVELOP A CLEARLY FOCUSED STRATEGY. WHAT IT TAKES IS A STRONG UNDERSTANDING OF THE BUSINESS ENVIRONMENT, COUPLED WITH A FULL APPRECIATION OF THE COMPANY'S CORE COMPETENCIES. WHERE DOES THIS SORT OF KNOWLEDGE COME FROM? IN BURGESS'S EXPERIENCE, "IF YOU WANT TO RUN A COMPANY, PRODUCT MARKETING IS PROBABLY THE BEST HOME DISCIPLINE. BECAUSE YOU DO REALLY GET A DEEP UNDERSTANDING OF THE TECHNOLOGY, AND OF THE MARKETS, AND OF THE OVERALL BUSINESS PROPOSITION."**

products that catered to the specific needs of the entertainment and design segments of the market. Within just a couple of years, Alias's software was favoured by industrial designers at virtually every major auto manufacturer and games manufacturer; it also helped create Hollywood blockbusters like *Jurassic Park, Forrest Gump* and *The Mask.*

When Burgess joined Macromedia, in 1996, it too was suffering from a lack of focus. The company had been successful at developing multimedia tools for the CD-ROM market, but this market was disappearing. Steven Frankel, managing director at investment banking firm Adams, Harkness & Hill, says, "Macromedia didn't have a vision. It was headed toward mediocrity. A lot of software companies get stuck in limbo when they don't generate any excitement or growth. But then Burgess came on and totally refocused the company on the Web."

Refocusing the company meant making a number of big changes. Ten percent of the staff was laid off. Seven of ten products in development were cancelled because it wasn't clear how they would generate revenues for Macromedia in its new market space. Burgess knew that if the company was going to grow, these changes were necessary. But recognising the need for change and actually moving into new market space are two different things, as Burgess observes. "When there's a new opportunity, companies want to go after it, but they also want to keep doing everything that they were doing before. It's hard to stop what you were doing."

Macromedia was able to stop what it was doing, and during Burgess's tenure has grown fourfold, from a $100 million to a $400 million company. Macromedia products are now market leaders in four areas on the Internet: designing websites (Dreamweaver, Fireworks and Flash); developing interactive Web content (Dreamweaver UltraDev and Director); delivering Internet applications (ColdFusion); and displaying rich content on any Internet access device (98.3 percent of Web users have installed Macromedia's Flash player). As the company's website says, "Macromedia is helping to define what the Web can be." Burgess thinks "the Internet is a natural place to continue to focus. The Internet is still a very nascent place. Most of the things that have been invested in, from an Internet perspective, don't work, which makes sense because people have only really been at it for three or four years. You can't make complex systems in that length of time. So there's really a great opportunity there."

Burgess makes it look easy for a company to develop a clearly focused strategy. What it takes is a strong understanding of the business environment, coupled with a full appreciation of the company's core competencies. Where does this sort of knowledge come from? In Burgess's experience, "if you want to run a company, product marketing is probably the best home discipline. Because you do really get a deep understanding of the technology, and of the markets, and of the overall business proposition."

When Burgess talks about "core competencies" he is alluding to the concept defined by management gurus C. K. Prahalad and Gary Hamel, first in their 1990 *Harvard Business Review* article and later in their 1994 bestseller, *Competing for the Future.* According to Prahalad and Hamel, a core competency is the "bundle of skills and technologies that enables a company to provide a particular benefit to the customer." Core competencies are difficult for competitors to imitate.

> **"DO GREAT WORK FOR YOUR CUSTOMERS. IF YOU TRY TO DO GREAT WORK FOR YOUR CUSTOMERS, THEN YOU'LL ALWAYS HAVE A LOT OF OPPORTUNITIES. IF ANYBODY'S INVESTED WITH ME, OR GIVEN ME AN ORDER OR ANYTHING, I JUST TRY TO EAT PAVEMENT TO MAKE THEM SUCCESSFUL."**

In a study of core competencies in multinational companies, published in the *California Management Review*, authors Mascarenhas, Baveja, and Jamil observed that "leading companies do not stand still and rest on their traditional competencies. Instead they develop new competencies that respond to or anticipate emerging business conditions." In Burgess's view, few Internet companies have done a good job of anticipating emerging business conditions on the Net. "I don't think the last two or three years have necessarily been the model for people to look at, as far as Internet businesses are concerned, because everybody kind of went nuts. A lot of it was fuelled by the venture capital community—sort of a gold rush down here that led to the funding of a lot of things that really shouldn't have been funded—it was business model du jour."

Burgess emphasizes that in the technology business it is important to be nimble. So companies should strive for focused strategies, and be prepared to change their focus if the business conditions change. Burgess thinks that this is a good prescription for countries as well. "Focus is just as important for a country as it is for a company. You know, this notion of core competence feeds on itself. You can't compete in every technology area. Canada isn't overnight going to become a leader in databases and a leader in telecommunications and a leader in nano-technology."

But Canada has been a leader in developing high-performance 3-D graphics for an interesting reason, as Burgess explains. "In the 1980s, Canada had basically all the leading companies in the world. They specialized in high-performance 3-D graphics right across the country. I like to think that's because it actually is rocket science that has to go into the software and the fact is, Canada doesn't have a rocket science industry. So those smart students—the ones who want to create new physics—since there's no military/industrial complex, they did it virtually, and a lot of that smart math occurred." Burgess does admit that 3-D graphics "didn't end up being a huge industry," perhaps because neither the companies nor the country adapted well as business conditions changed.

There's one more important lesson to learn from Burgess. He knows he's not the first to give this advice, but observes, "Just because it's not unique, doesn't make it untrue." Companies can't go wrong if they invest in customer service. "Do great work for your customers. If you try to do great work for your customers, then you'll always have a lot of opportunities. If anybody's invested with me, or given me an order or anything, I just try to eat pavement to make them successful. And you shouldn't be afraid of making mistakes, or admitting them, but if you go the last mile for your customer, good things seem to happen; people like doing business with you."

John Eckert Paul Chen_

SURVIVING THE LONG WALK THROUGH THE NIGHT

INNOVATOR_ PAUL CHEN AND JOHN ECKERT, SERIAL ENTREPRENEURS, COACHES, FINANCIERS
INNOVATION_ DEVELOPED A WORLD-CLASS E-MAIL MANAGEMENT AND DELIVERY FIRM, WHICH WAS SOLD TO INDUSTRY LEADER DOUBLECLICK

The names Media Synergy, FloNetwork, and McLean Watson Capital might be familiar to business students. That's because these companies are the subject of three Harvard Business School cases. The cases begin with Media Synergy's founder, Paul Chen, negotiating venture financing with McLean Watson Capital. After round two of financing (discussed in the second case), Media Synergy changed its focus, and its name, rebranding itself as FloNetwork. The third case in the series looks at FloNetwork's acquisition by DoubleClick, completed in 2001.

Chen now works with McLean Watson Capital co-founders John Eckert and Loudon Owen as an entrepreneur-in-residence at the venture capital firm. When asked recently if, as a potential outside investor, he would have extended financing to Media Synergy in the early days, Chen said, "No, the company was so inexperienced. There was passion and commitment, but for me to put my own money into it would have been scary." Of course, Chen did put in his own money. He and his sister Mina poured all their energy into the company to get it started, but it was Owen and Eckert who stepped in as the early investors. What was it about this inexperienced brother–sister team and their small startup that was of interest to McLean Watson Capital?

As Owen explains, "When we met them, they had that element that you cannot put in a bottle and you cannot adequately describe. It was drive and integrity and insight. Something was going to happen in the organization." And although Chen suggested that the company was inexperienced, it had developed a product (low-end 3-D animation software) that was on sale at U.S. retailers like Egghead and CompUSA. Eckert says that Chen had "the experience of success. Getting a product on the shelf was a huge achievement. So many people start out trying to do that, and so few get there."

But the company needed help developing additional marketing channels, and was looking for capital. Chen remembers meeting with "at least 20 or 30" VCs. He felt the best rapport with McLean Watson Capital. Several years earlier, Eckert and Owen had financed Softimage, the 3-D animation company founded by Daniel Langlois. They helped take the company public with a NASDAQ IPO and served as joint COOs for the company until it was sold to Microsoft in 1994. The company's software "is still the software of choice for all the high-end 3-D animation work that you see coming out of Hollywood." Owen and Eckert were very familiar with the graphics industry and decided to invest in Media Synergy.

Venture capital deals are set out in term sheets. Negotiated between the company seeking funding and its venture investors, a term sheet typically outlines the capital structure of the deal and the shareholders' agreement. But as a legal document, the term sheet doesn't come close to capturing the spirit of a successful partnership among a VC and an entrepreneur. For Eckert and Owen, a deal is really about investing in people. Eckert notes that it is McLean Watson Capital's style to back "enthusiasm and commitment, rather than more experience with a lack of passion." He says, "When we made the investment in Media Synergy, the thing that made us say, 'Let's go ahead,' was the belief that this was going to be a fun opportunity whatever happened. That was when we said, 'Okay, let's do it. Let's put the money in.'"

Owen and Eckert believed that investing in Chen's company was going to be successful because the company had a very special culture. In fact, Eckert suspects that Chen

IT WAS A VERY YOUNG COMPANY. IT WAS EXTREMELY DRIVEN AND DEDICATED. AND IT WAS INTERESTING BECAUSE IT WAS EXTREMELY MULTICULTURAL; IT WAS NOT HOMOGENEOUS IN ANY SENSE OF THE WORD. FLO WAS THE SORT OF PLACE WHERE PEOPLE REALLY BELIEVED IN WHAT THEY WERE DOING, AND WENT TO EXTRAORDINARY LENGTHS TO MEET A DEADLINE, OR MEET THE EXPECTATIONS OF A CUSTOMER.

"doesn't even know how special it was," although he might be able to see it if he looks back 10 years from now. "Culture is very hard to define on paper. It's not like you can put down a 10-point plan for getting it. It's a chemical compound. You can try everything, but still may never be able to create that special culture. FloNetwork had it, and to tamper with it would have, I think, destroyed it."

Although it's not easy to set out a recipe for a strong corporate culture, Owen recognises that investing in the right people is key. "People drive everything. You look for drive, integrity, brains, and courage." The most successful people that McLean Watson Capital has worked with "are individuals who care about what they do. It's never just about the money, ever. It's always about the vision, the dream, and what they want to accomplish and achieve. It's genuine." Paul Chen is one of these people. "He's very soft-spoken and he's very humble, but he's absolutely wonderful at instilling genuine confidence in people, and working with them. And they will work so hard for him it's unbelievable."

In Chen's case, this meant that his team stuck with him through the period Eckert calls "the long walk through the night." At the time of McLean Watson Capital's initial investment, Media Synergy was struggling to extend the market for its shrink-wrapped software product. In the fall of 1996, the company launched a new product, a player that allowed people to send multimedia e-mails. Half a million copies were sold, and Hallmark used the software to deliver electronic greetings. But there were problems with the technology, and Media Synergy discovered it was simply too expensive for a small company to market a single product to consumers. Eckert says, "By 1997, it was clear that this wasn't going to make us rich. So there was about a year of real soul-searching, and trying to figure out what to do. Money was getting tight."

The company decided a business-to-business focus was more appropriate, and negotiated a partnership with RealNetworks to integrate Media Synergy's multimedia player into its next release. The deal would bring a steady revenue flow to the company, but it required at least a year of development work. At the same time, the company was working on a server technology that would allow businesses to send mass e-mails. Chen recalls, "We spent about ten months working on the RealNetworks project before we decided it was going nowhere and just cut it off." Focus shifted to the server technology, which was ready to ship as a software product in late 1998. In early 1999 (Chen remembers the exact date), a decision was made to shift away from a product focus to an application service provider model, where the company would sell their technology as a service.

Many things began to happen then. Paul made the difficult decision to step back in favour of Eric Goodwin. Eric was a serial entrepreneur from Ottawa who also understood what it took to grow a company quickly. He was able to institute the controls and strategic planning required while preserving and even improving the Flo culture. Flo quickly grew from 38 employees in March 1999 to 139 one year later. After one year, the company had finally found a winning combination. It rebranded itself as FloNetwork, and in 2001, when it was sold to DoubleClick, Flo had 210 employees and a blue chip list of customers, including Procter & Gamble, Continental Airlines, and Barnes & Noble.

Chen's team remained committed throughout this period. Eckert observes, "It was a very young company. It was extremely driven and dedicated. And it was interesting because it was extremely multicultural; it was not homogeneous in any sense of the word. Flo was the sort of place where people really believed in what they were doing, and went to extraordinary lengths to meet a deadline, or meet the expectations of a customer."

When asked about the lessons he learned developing FloNetwork, Chen returns to the importance of culture. "One of the lessons I learned is that strategy, for a company, is very important. But a lot of times people ignore the fact that strategy needs to be executed by people, so they ignore their people, and the culture of people becomes secondary. In fact, strategy and people have to fit. At Flo we were lucky, because we sort of subconsciously developed our culture in a way that fit our strategy. But next time—and I know I'm going to build another company—next time I'll be paying equal attention to the cultural aspect."

The FloNetwork culture was strengthened by "a lot of fun company get-togethers. It might have been work-related, but there was always a strong element of fun." But these events weren't organized by management. Chen remembers that "a lot of them were very spontaneous. They weren't organized because somebody at the top said, 'We should do this'; they would just happen." Eckert continues, "People did not want to miss Flo events. They would drive in, fly in. They'd come in even if they were sick. They weren't things that you had to drag them kicking and screaming to."

The FloNetwork team is no longer together. Some employees went to DoubleClick; others went off to new ventures. But the Flo spirit lives on. Says Eckert, "They all want to come back. They all yearn for the old days and love to get together for Flo reunion parties."

Shane Chrapko Val Pappes_ Evan Chrapko_

"SEED" FINANCING

INNOVATOR_ EVAN CHRAPKO, SERIAL ENTREPRENEUR, COACH, FINANCIER INNOVATION_ CO-FOUNDED THE DOCSPACE COMPANY, AND SOLD IT TO CRITICAL PATH ONLY 19 MONTHS AFTER LAUNCH FOR $568 MILLION (U.S.)

Evan Chrapko often gets asked, "How do you get from a hog farm in the middle of nowhere to Canada's biggest pure Internet deal?" The self-made multimillionaire doesn't mind the curiosity, since he attributes much of his business success to his rural upbringing. Many of the lessons he learned on the farm can be applied to business. For Chrapko, growing up as an Alberta farmboy gave him a "single-minded focus on sacrifice, ethics, and self-sufficiency," qualities that catapulted him from an isolated, self-conscious kid to the articulate, confident, and highly competent CEO he is today.

Chrapko fondly recalls how he kept a model of a Lamborghini on his nightstand to serve as a symbol of how he would someday leave the farm behind and become one of Canada's leading technology entrepreneurs. The childhood dream became a reality in 2000, when Chrapko's startup, the DocSpace Company (an Internet-based firm that electronically stores, shares, and transfers confidential documents) was sold to Silicon Valley-based Critical Path for $568 million (U.S.). In fact, the high-profile deal would make the record books as a project that went from launch to sale in just 19 months.

> **AS THE COMPANY'S LEADER, CHRAPKO ALSO PUSHED TRUST AND RELIANCE TO THE FOREFRONT, AND WOULDN'T TOLERATE ANY EMPLOYEE WHO LACKED BASIC INTEGRITY AND TEAMWORK SKILLS. HIS REASONING WAS SIMPLE: "ON THE FARM, YOU CAN BE PUT IN MORTAL PHYSICAL DANGER IF THE PERSON AHEAD OF YOU DOES SOMETHING SLOPPY OR STUPID, OR FAILS TO COMMUNICATE THE WHOLE."**

Although it might seem like just another dot-com overnight success story, in reality, Chrapko took 15 years to prepare for that shining moment. As he points out, forecasting for the future is inherent in farm life. "By nature you plan decades ahead, not just what you're having for dinner that day." His entrepreneurial strategy was mapped out while sitting day after day in a tractor cab in the fields of his family farm near Brosseau, Alberta. The plan included four years in business school at the University of Alberta, three years to earn his accountancy designation, and then another three years at Columbia University's law school. Getting the proper accreditation gave him the confidence he needed to believe in himself as an entrepreneur. "I felt really self-conscious about who I was, a kid in rubber boots slopping through the pig shit, to think that I could end up at a law school, let alone an Ivy League law school."

It was some time during his final schooling efforts that he found the time to turn around a near-bankrupt Edmonton-based company. This challenge piqued his interest for saving companies from the brink of financial failure. Chrapko put his entrepreneurial ambition to the test by helping Cel Corp., which had a solid technology product but was marred by poor marketing efforts and major managerial problems. He recruited his younger brother Shane to head up sales, and soon the two brothers had made good on the promise of profitability.

Buoyed by that success, the brothers hooked up with two family friends, Valerian Pappes and Cristy Rowe, to form A-LIVE Holdings, a consulting firm set up to rescue companies from financial ruin. Then Cyberian Networking Corp., a Vancouver-based Internet service provider, asked Chrapko to help with one of their new products. Chrapko and his partners were intrigued with Cyberian's technology idea, but were convinced that it had to be developed as a service business—not just a product. A-LIVE Holdings bought the technology in 1997 and started the DocSpace Company along with Sandra Wear, Michael Serbinas, and Daniel Leibu.

Relocating the company to Toronto's technology hub, the founders and several early employees shared accommodation and workspace in a low-income rental house. The partners ran the business like a family farm "that lived, worked, and played together." Like most startups, DocSpace faced a scarcity of resources, but the farm-bred brothers solved the problem by relying on the land for sustenance, despite their new location in the heart of downtown Toronto. "The lack of resources paralleled exactly what you deal with on the farm every single day, right down to our having to plant a garden out back. We needed to augment our diet of Kraft Dinner and so we threw some seeds in the ground, and that was nothing unusual to us."

As the company's leader, Chrapko also pushed trust and reliance to the forefront, and wouldn't tolerate any employee who lacked basic integrity and teamwork skills. His reasoning was simple: "On the farm, you can be put in mortal physical danger if the person ahead of you does something sloppy or stupid, or fails to communicate the whole."

Chrapko is somewhat surprised by how often his own integrity was put to the test during the growth of the business. Although many investors were interested in funding and acquiring his valuable company, Chrapko says he took the time to find the "right deal with the right partner," rather than accepting anyone who was willing to give him the money he desperately needed to drive the business forward. "I refused money from lots of different offerers along the way, because their terms were nothing short of exploitative. Some potential acquirers saw how we were living and assumed that we would just jump at anything because they saw our conditions over at the house—the dumpster furniture and so on."

Chrapko's ability to hold out for a deal that was the best fit for his company resulted in a huge payoff and made millionaires out of most of his employees. The lesson

> **ALONG WITH HOLDING STEADFAST TO A GOOD REPUTATION, CHRAPKO ADVISES FLEDGLING ENTREPRENEURS TO AVOID BLIND EXPECTATIONS FOR THEIR BUSINESS, ESPECIALLY IN THE FAST-PACED WORLD OF TECHNOLOGY. "YOU ABSOLUTELY HAVE TO HAVE A VISION, BUT IT'S GOT TO BE BIG, ESPECIALLY IN TECHNOLOGY. IT'S GOT TO BE WORLD-SCALE, OR WORLD-CLASS IN SCOPE, AND YOU'VE GOT TO BE MOVING TOWARD IT AS FAST AS YOU HUMANLY CAN. AND WITH QUALITY, NOT SOME SLAP-DASH NONSENSE EXERCISES."**

he now passes on to the next generation of entrepreneurs is to "stick to the high road. Don't sell out, because you only have one reputation. You are born with it. We got ours given, and buffed up, on the farm and we didn't want to lose it, especially not for someone else who could care less about whether or not we survived the next day, or the economic downturn or whatever. Remember, they're in it for themselves."

Along with holding steadfast to a good reputation, Chrapko advises fledgling entrepreneurs to avoid blind expectations for their business, especially in the fast-paced world of technology. "You absolutely have to have a vision, but it's got to be big, especially in technology. It's got to be world-scale, or world-class in scope, and you've got to be moving toward it as fast as you humanly can. And with quality, not some slap-dash nonsense exercises."

Now, having realized his own vision, the entrepreneur has set up the Evan V. Chrapko Foundation to help inspire a new generation of young Canadian talent to realize their dreams, regardless of how unattainable they may appear. "I have a lot of time for something that's going to reach out to youngsters. In the back of my mind, I am thinking if half of one of my words lights up some kind of spark or fire in a fisherman's son, or an Inuit daughter, or someone else sitting in a tractor cab in the middle of nowhere, who doesn't know whether they have the right to think that they can do it, too, I need them to know that they can."

At age 36, few would deny that Evan Chrapko has reached the pinnacle of entrepreneurial success, but what of his own boyhood dream to one day own a Lamborghini? You guessed right. The real life version is now sitting in his driveway. Chrapko admits it stands there "in honour of that little kid who had a dream. Because if you don't execute on it, then what the hell's the point of carrying on the dream all that time?"

SUCCESS AT $26.95

INNOVATOR_ DR. GURURAJ "DESH" DESHPANDE, SERIAL ENTREPRENEUR, ACADEMIC BRIDGE

INNOVATION_ BUILT ONE OF MOST REVERED AND SUCCESSFUL NETWORKING COMPANIES IN THE WORLD

Deshpande netted $26.95 from the sale of his first company. Seven years later, in 1997, his second company sold for $3.7 billion (U.S.). His third company broke IPO records when it was listed on NASDAQ in October 1999. Along the way his companies gained considerable recognition: *Fortune* magazine's "Top 25 Very Cool Companies of 1996," "Cool Companies for 1999," and "Top Ten Picks for 2000." *Red Herring* named Sycamore a "Top 100 Company of the Electronic Economy." Deshpande's products also received accolades—*Telecommunication* magazine's "Top 10 Hottest Technologies," and "2000 Product of the Year" from *Network* magazine. Clearly Deshpande has some stories to tell about high-tech startups.

Deshpande came to Canada from India, where he had gained a B.S. in engineering from the Indian Institute of Technology. Offered a scholarship by the University of New Brunswick, he completed an M.E. in electrical engineering, and then went on to garner a Ph.D. in data communications from Queen's University in Ontario. He might have stayed as a professor in the pleasant little town of Kingston, Ontario. But one of his colleagues, Dr. Peter Bracket, persuaded him to join him in a move to Toronto to grow Codex Corp., a small subsidiary of Motorola. And here Deshpande found that he liked the excitement of a startup. Although he was the junior member of the management team, he discovered that growing a business from 20 to 400 people and $100 million in sales was exciting and rewarding. In six months he went from writing code to running a team of 30 or 40 engineers. What insight did he gain from that experience? He says, "You understand the technology, so you know what technology can do, but you also find out what people need. If you can do something that is simple and very cost-effective, but has a huge value to somebody, then, bang, you've got a business!"

In 1984, recognising that venture capital was scarce in Canada, he went to a major source of high-tech money—Boston. Never mind that he knew nobody, that he had no capital. He worked for awhile, and then started his first solo company—Coral Network Corp.

His advice to budding entrepreneurs: "Save enough money to cover the first 15 to 18 months and live frugally—live like a student!" Coral didn't quite work out and Deshpande sold the company for $26.95. But what it did teach him was that while staying the course is important, so is the recognition that you can and should walk away from some of your own creations—either when you need to cut your losses or when you are not needed anymore.

His next venture was based on a deceptively simple concept: "Just like every phone is connected to every other phone, every computer had to connect to every other computer." So he decided that the market opportunity was to build data networks for public carriers. He started Cascade Communications in his basement and raised $125,000 in venture capital. But Deshpande realized he needed a seasoned CEO and hired Dan Smith, an engineer and Harvard MBA. The pair had "instant chemistry," he says, and together they managed to make Cascade switches a mainstay of the rapidly growing high-speed networks. Within six years, Cascade had grown to a 900-person business with $500 million in sales. They sold the company to Ascend Communications for $3.7 billion (U.S.) in 1997. (Lucent then bought it in 1999 for nearly $20 billion.) The success of Cascade allowed Deshpande a break to reflect on his next venture. His advice to other serial entrepreneurs is always to take a long break to focus on the next big idea. Or, as he puts it, "to dream."

HIS ADVICE TO BUDDING ENTREPRENEURS: "SAVE ENOUGH MONEY TO COVER THE FIRST 15 TO 18 MONTHS AND LIVE FRUGALLY—LIVE LIKE A STUDENT!" CORAL DIDN'T QUITE WORK OUT—HENCE THAT $26.95 CHEQUE. BUT WHAT IT DID TEACH HIM WAS THAT WHILE STAYING THE COURSE IS IMPORTANT, SO IS THE RECOGNITION THAT YOU CAN AND SHOULD WALK AWAY FROM SOME OF YOUR OWN CREATIONS—EITHER WHEN YOU NEED TO CUT YOUR LOSSES OR WHEN YOU ARE NOT NEEDED ANYMORE.

During his break in 1997, when he took time to mentor and invest in Web start-ups in the Boston area, he came to see that if Cascade was about making effective use of bandwidth, his next venture would take advantage of the growing demand for bandwidth in the networked economy. Matrix partners, which had provided capital for Cascade, introduced him to Richard Barry and Eric Swanson, two optical networking experts from MIT with ideas on how to improve the efficiency of high-speed networks. In 1998 Sycamore Networks was born.

Sycamore's focus is optical switching. Its products provide dynamic bandwidth allocation and network management to enable networks to handle data more efficiently and in response to variable demands. Its IPO in 1999 saw an initial offering price of $38 close at $185. This initial success gave the founders the capital to invest aggressively in infrastructure and international expansion. Rapid product development and technological innovation helped catapult Sycamore into the big leagues in an unforgiving marketplace. Analysts have noted that the company's "software has cut the time it takes to expand or contract bandwidth from six days to six minutes." The company was touted as a formidable rival to telecom giants Nortel Networks, Cisco Systems, and Lucent Technologies and has experienced phenomenal growth.

> BUT DESHPANDE IS UNFAZED. HE POINTS OUT THAT EVERY NEW INDUSTRY FACES TURBULENCE AND THAT THE HIGH-TECH INDUSTRIES HAVE DONE MORE IN THE LAST THREE OR FOUR YEARS THAN THE "OLD ECONOMY" COMPANIES ACCOMPLISHED IN 20 YEARS. HE FIRMLY BELIEVES THAT INNOVATION IN TECHNOLOGY OVER THE NEXT DECADE WILL MEAN CONTINUED AND DRAMATIC CHANGE. HE'S ALSO A REAL BELIEVER IN MARKET FORCES, SAYING, "LET THE WINNERS WIN AND THE LOSERS LOSE."

The company was dubbed by *Canadian Business* Magazine in 2000 as "the money tree," with Deshpande as its $5 billion man. *India Today* christened him "the world's richest Indian." But the money itself does not appear to interest him. "Stock prices go up and down. It's just a number, so that's not a really good goal to shoot for. Trying to be number 1 [in wealth] is not enough to sustain one's initiative."

Since those heady days, Sycamore's stock price and performance have shared the hard times with the rest of the telecom industry. But Deshpande is unfazed. He points out that every new industry faces turbulence and that the high-tech industries have done more in the last three or four years than the "old economy" companies accomplished in 20 years. He firmly believes that innovation in technology over the next decade will mean continued and dramatic change. He's also a real believer in market forces, saying, "Let the winners win and the losers lose."

Despite the challenges of downsizing and cost cutting, Deshpande's initial vision is intact. He says that bandwidth is a valuable resource. The Internet is core to virtually every business today and to 25 percent of the world's population, whereas "ten years ago, the Internet was relevant only to a bunch of MIS guys." Optical technology remains a key to improving capacity and productivity. But it's important to take a long-term perspective. Investments now will yield benefits down the road. The current problem is in part a misalignment between pricing and value. He acknowledges the challenge but maintains, "To me it's like playing a game. The game gets harder but it doesn't make it less interesting."[1]

Deshpande still looks back fondly on his alma mater and his adopted country—Canada. Like many others, he emphasizes the value of Canada's education system. "It's extremely disciplined. They're very demanding and crank out very good professionals." He also sees a different environment for startups today than in 1984, when the lack of venture capital drove him to Boston. He maintains that entrepreneurial culture rather than location is a critical ingredient for success in building a company. "Number one, you need the innovation; you need something new. And then you need a great bunch of people. And then you need a really good market." And for him that market is global.

To Deshpande, "Life is a series of accidents." But his series of entrepreneurial successes seems to be anything but accidental. He combines depth in technology with solid business instincts. He's someone who can talk about ethics in business and cite Gandhi as a role model in the same sentence as he talks about Bill Gates.

Deshpande is also as proud of his teaching and coaching as he is of his business success. He still expresses a love of teaching and remembers fondly winning a teaching award as a professor. He aims to mentor a startup a year and has committed several days a year to MIT, where he is on the board and has funded the launch of the Deshpande Centre for Technological Innovation—intended to help young companies to grow in partnership with leading universities.

And where is that cheque for $26.95? A copy of it is framed on his wall at the urging of his wife, Jaishree, to remind him that success is never guaranteed!

---

[1] Joe McGarvey, "Sycamore Networks: The Road Ahead," *The Net Economy*, January 2002.

THE ORIGINAL
SURFER

INNOVATOR_ BRIAN EDWARDS, SERIAL ENTREPRENEUR INNOVATION_ SOLD HIS ORIGINAL STARTUP TO BELL AND WENT ON TO FOUND ONE OF CANADA'S MOST SUCCESSFUL AND TALKED ABOUT E-COMMERCE COMPANIES—BCE EMERGIS

Brian Edwards doesn't talk like a technology guru. In fact, he had already clocked more than 20 years' experience when the current *wunderkind* crop were still in diapers. But Edwards's combination of solid business smarts and an opportunistic eye have enabled him to successfully catch and ride every major technology wave, from electronic messaging to EDI, to the Internet and B2B applications. BCE Emergis, which was created by joining Edwards's company MPACT Immedia with Bell Canada's messaging divisions in 1997, passed the half-billion-dollar mark in 2000, making it the largest e-business venture in the country. Although Emergis is facing challenges, including IBM's recent entry into its markets, that does not detract from its impressive growth and unparalleled success in the U.S. market. Edwards decided to move on in early 2002, but expects his successors to take Emergis "to a new level."

Although every venture the serial entrepreneur has touched seems to have turned to gold, Edwards says he simply relies on the business basics of good marketing and organizational skills. When accused of being smart, he says, "Smart or lucky, I don't know." But a walk down memory lane with Edwards shows how the astute application of basic business skills, coupled with the instinct for seeing and seizing opportunities before anyone else, allowed him to ride the waves of technology for more than 30 years.

Edwards was no kid tinkering with games in a garage. He spent more than 15 years working with big companies—Burroughs Corporation, then IST—before many of the young turks were even born. He gained extensive experience in sales and management and engineered a major turnaround before deciding he wanted equity in a company of his own.

But he did not have a lot of money. So he parlayed his management smarts into shares in MPACT Immedia, a company that targeted the emerging electronic messaging market at a time when few people even knew what the term meant. Edwards wanted to build a public service for electronic messaging, which he believed would become a huge market. He is hard pressed to articulate the combination of factors that led to the insight. "You see what the business does, how it works, and that it makes sense. But when we went out and talked to people, especially investors, they would say to me, 'Why do I need this stuff? I have a fax.' At that time no one conceived that someday everybody would get connected, and everybody would use electronic messaging. But we believed it. From the very beginning of MPACT, every person had a computer, and got plugged into the network." Edwards's instincts paid off—he took the company from $700,000 in revenues and $1 million in losses, through a public offering, which raised $2.5 million, to revenues of $3 million by 1993.

At that time, the company's service was running on Tandem computers (now Compaq) and Edwards sat on a Tandem advisory committee board. "One of the things that Tandem was saying was, 'What do we have to do to be a better partner?' The software companies were arguing, 'Well, you have to stop competing with us. So, if you're in the software business, get out of it.' I thought that was interesting, so I called the CEO of Tandem and told him I'd be interested in buying his EDI business." Just like that: Edwards saw something, he grabbed it and caught another wave—the exploding business of electronic data interchange.

> EDWARDS SAYS THE LESSONS ARE DEAD SIMPLE: "FIRST AND FOREMOST THESE ARE BUSINESSES, AND SO PEOPLE HAVE TO RUN THEM AS BUSINESSES. YOU MAY NEED TECHNOLOGY TO SUCCEED, BUT IF YOU GET SO WRAPPED UP IN THE TECHNOLOGY THAT YOU FORGET ABOUT THE BUSINESS, IT'S NOT GOING TO WORK."

"We had unstructured messaging (e-mail and text) and we had added to that structured messaging (electronic data interchange)." By adding Tandem's business and sales force, MPACT became an $8 million business. "But they were losing a ton of money, so we had to reduce costs and headcount in order to grow."

By 1997, the Internet had emerged as a big threat to EDI and once again Edwards spied an opportunity. "TotalNet, a Montreal-based Internet Service Provider (ISP), was on the verge of making money, but nobody wanted to invest in it because the other two public offerings (iStar and Hookup) had been a disaster. So we took a risk and invested a few million dollars in MPACT shares. At the very least, we needed to understand the Internet better, to try to make the conversion from EDI to the Internet. We ended up selling TotalNet at the end of '99 for about five or six times what we had paid in cash."

LIKE MANY OTHERS, EDWARDS SEES CANADA'S COMPETITIVE ADVANTAGE NOT JUST IN TERMS OF TECHNOLOGICAL INFRASTRUCTURE: "CANADA HAS GOT SOME VERY GOOD SKILLS, AT A MUCH CHEAPER RATE THAN THE AMERICANS. IT'S VERY EXPENSIVE TO DO BUSINESS IN THE UNITED STATES. PEOPLE COST MORE; BENEFITS COST A LOT MORE. PUTTING UP OF DATA CENTRES AND PRODUCTS, THEY COST AS MUCH OR MORE. WE ARE WELL POSITIONED IN THOSE MARKETS."

Once again, Edwards had spied the future and was positioning for the next wave. This time, it wasn't a technology; it was a scale of magnitude. "We were really keeping our eye on our big competitors. In 1997, Harbinger, a company about our size, had done a couple of deals that took them from the $25 million range to nearly $100 million. And so we said to ourselves, 'We're not going to be able to survive at the size we're at. We need to find a partner that is a good strategic fit with our business.' Bell was our major competitor in Canada. Its business was about twice the size of the MPACT business, but we were making money and Bell wasn't. So we said to Jean Monty, 'Why don't you put your business together with ours in a public company environment? You get good value for the business, and we'll help to grow it. You're twice our size, so we'll give you twice as many shares as we've currently got outstanding, so you'll also get control. If you don't like it, you can sell us off. If you like it, you could eventually buy 100 percent if you wanted to.' When we put the two businesses together in September '98, we had about $75 million revenue. In '99 we moved the business to $187 million, and then to $465 million in 2000, and to $656 million in 2001. Putting together the resources of MPACT and Bell, with the Bell brand, ended up being a huge win for everybody."

Edwards says the lessons are dead simple: "First and foremost these are businesses, and so people have to run them as businesses. You may need technology to succeed, but if you get so wrapped up in the technology that you forget about the business, it's not going to work. It might work for a little while—for example, during the whole Internet craze, when the market got so excited about the Internet—but it's not something that's sustainable. Managing the cash is one important thing." He also stresses that while you keep a watchful eye on what you are doing, you also have to keep scanning for the next opportunity, particularly the opportunities that will move you to the next level. "The EDI deal with Tandem brought us to another level. The Internet deal with TotalNet brought us to another level. Certainly the Bell deal brought us to another level. The Bell deal drove a tremendous amount of momentum around Emergis."

Edwards is also clear on another point: "To be a successful Canadian company, we have to be able to win in the U.S. The U.S. is the world's most competitive technology market, certainly, and it's no use getting to be a big Canadian player, because by the time the Americans get to be big, they are 10 times as big as we are, if they go about the same thing proportionately."

Like many others, Edwards sees Canada's competitive advantage not just in terms of technological infrastructure: "Canada has got some very good skills, at a much cheaper rate than the Americans. It's very expensive to do business in the United States. People cost more; benefits cost a lot more. Putting up of data centres and products, they cost as much or more. We are well positioned in those markets."

The way to be successful in this new world will be to anticipate the next wave of applications. "So the questions are 'What are the new functions? What are the new technologies? What are the new applications that will reside right in the network?'" The jury may still be out on BCE Emergis's prospects, but there is little doubt that Edwards is currently looking for the next wave to catch.

NORM FRANCIS "It is critical to be able to recognise a trend that is going to be significant and one that you can build businesses on. That's a differentiator. Obviously you have to start with an excellent opportunity. But you also need to be able to build a company around it. You need a combination of the strategic and the tactical."

THE GRAND
SLALOM

INNOVATOR_ NORM FRANCIS, SERIAL ENTREPRENEUR INNOVATION_ STARTUP PIONEER WHO SET THE INDUSTRY STANDARD IN BOTH ACCOUNTING AND CUSTOMER RELATIONSHIP MANAGEMENT SOFTWARE

The path to high-tech riches is full of unexpected curves and obstacles that can send you flying off course, according to software tycoon and serial entrepreneur Norm Francis. While "the vision thing" is important, so is flexibility. "You shouldn't let yourself become too attached to your corporate vision."[1] The world is constantly changing and even the best-laid plans can go astray. The key is being able to adapt, "like a skier racing in the giant slalom." Francis should know. His first startup, Basic Software Group (BSG), created in 1979, developed AccPac, which rapidly became the gold standard in accounting software and was acquired by Computer Associates (CA) a few years later. In 1994, Francis and partner Keith Wales founded Pen Magic Software, which evolved into Pivotal. Pivotal's customer relationship management solutions are now sold in 35 countries; the company is worth more than $200 million.

---

[1] Peter Caufield, "Successful Companies Constantly Adjusting" *Computing Canada,* February 19, 1999.

According to Francis, successful technology companies may develop strategic "intents" or goals, but they have to be able to make "micro adjustments" to their strategies as they go along. He likes to draw an analogy to skiing. "Good skiers visualize in their minds the whole race, including crossing the finishing line, before they start. Without such an image in your head, it's hard to make the right moves and adjustments as you're racing down the course, over the bumps, around the turns and through the gates."[2]

> **PIVOTAL WAS ONE OF THE FIRST SOFTWARE VENDORS TO UNDERSTAND THAT LINKS TO CUSTOMERS WERE POTENTIALLY THE MOST SIGNIFICANT OPPORTUNITY IN THE VALUE CHAIN. AND THAT DID NOT MEAN JUST AUTOMATING SALES; IT INCLUDED MANAGING INFORMATION ABOUT CUSTOMER NEEDS, MANAGING SALES, SERVICE, AND MARKETING.**

"In the case of AccPac, it was these new small computers at the time, microcomputers (later christened personal computers), that created the opportunity to computerize financial and accounting tasks for small- to medium-sized businesses which previously could not afford their own computers." Later, Pivotal was created to exploit the trend in the early '90s of companies choosing to focus on "a more one-to-one marketing approach, aimed at understanding and meeting their customer's individual needs." Indeed, Pivotal was one of the first software vendors to understand that links to customers were potentially the most significant opportunity in the value chain. And that did not mean just automating sales; it included managing information about customer needs, managing sales, service, and marketing.

[2] Peter Caufield, "Successful Companies Constantly Adjusting" *Computing Canada*, February 19, 1999.

When the explosion of the Internet changed the game, Pivotal redesigned its software to enable clients to co-ordinate Web-based selling and customer-relationship management, ensuring that appropriate messages were delivered to each customer and that orders were processed smoothly. The company has made strategic acquisitions to enable it to extend the reach of customer relationship management across the Internet.

> "SUCCESS BREEDS SUCCESS," SAYS FRANCIS, EXPLAINING HOW HE AND KEITH WALES USED THEIR TRACK RECORD WITH ACCPAC ACCOUNTING SOFTWARE TO ATTRACT SILICON VALLEY VENTURE CAPITAL GIANTS SUCH AS KLEINER PERKINS. "IT ADDS CREDIBILITY TO HAVE MAJOR-NAME INVESTORS AND WELL-CONNECTED BOARD MEMBERS," SAYS FRANCIS.

Like the other opportunities that Francis has grasped, the company moved into the e-business space earlier than most. "We were the first to offer a completely Internet-based solution for demand-chain management," says Francis. He also saw the potential of "the market segment of one," which drives the demand for tools that help companies customize their efforts to customers. Francis is known in the industry for the concept of "360-degree customer relationship management," which unifies the activities of an organization around the customer relationship to build customer loyalty. The trend he saw was "that change or shift in business behaviour. The opportunity was to create software that could help companies do a better job in dealing with their customers."

But moving to target big companies with a high ticket package (more than $200,000 U.S.) was a new game for Pivotal. "Success breeds success," says Francis, explaining how he and Keith Wales used their track record with AccPac accounting software to attract Silicon Valley venture capital giants such as Kleiner Perkins. "It adds credibility to have major-name investors and well-connected board members," says Francis. "It's easier to do a new venture where you are attracting employees, board members, customers, investors, if you have a track record. There's no question about it." When top-notch employees join a company, "they want to know they're joining a winner."

The company's location positioned it well to hook up with Silicon Valley. "We're at the north end of the West Coast technology strip: Vancouver, Seattle, Portland, San Francisco Bay area." And they had access to Canada's best and brightest. "We have been able to create

FRANCIS IS KNOWN IN THE INDUSTRY FOR THE CONCEPT OF "360-DEGREE CUSTOMER RELATIONSHIP MANAGEMENT," WHICH UNIFIES THE ACTIVITIES OF AN ORGANIZATION AROUND THE CUSTOMER RELATIONSHIP TO BUILD CUSTOMER LOYALTY. THE TREND HE SAW WAS "THAT CHANGE OR SHIFT IN BUSINESS BEHAVIOUR. THE OPPORTUNITY WAS TO CREATE SOFTWARE THAT COULD HELP COMPANIES DO A BETTER JOB IN DEALING WITH THEIR CUSTOMERS."

a very strong and stable software development operation headquartered in Vancouver. And I think that's a tribute to the kind of training that is provided at Canadian universities." In the immediate area alone, "there are three very strong computer science schools: UBC, SFU, and UVic. I think that has been a key building block of the company."

In fact, talent is so key to his business that Francis acknowledges "the biggest thing I've learned in building a business is that I would always hire more senior people earlier in the company. I think sometimes I've been too conservative in that regard. When you're smaller it's hard to attract really seasoned management." An entrepreneur needs to know "when it's time to stop being CEO. Entrepreneurs are sometimes much better at starting companies than at running them. Sometimes a CEO needs to seriously consider handing over the reins to someone who's better at managing the day-to-day affairs of the company."[3]

How does he do it? "It is critical to be able to recognise a trend that is going to be significant and one that you can build businesses on. That's a differentiator. Obviously you have to start with an excellent opportunity. But you also need to be able to build a company around it. You need a combination of the strategic and the tactical." He credits his combination of technology know-how with business skills. "I had a computer science degree, but I'm also a chartered accountant. I can understand technology, but I also can understand business." He can see the opportunities that technology provides for solving business problems, but he does not get romantically involved with the bells and whistles of technology for its own sake. "Business is business," he says.

And Francis is convinced there is nowhere to go but up. "We believe we're at the early stages of a very long and large market—the customer-facing side of businesses. Much of this relates to the merging of customer relationship management and e-business. There are many opportunities for companies to use the Internet to extend their businesses to incorporate their business partners, such as sellers or agents, on the customer side of the business and to connect in better ways with customers themselves. But Francis is keenly aware of how quickly markets and competitors can change. And, of course, as a skier you want to keep your eyes open or you are dead.

[3] Peter Caufield, "Successful Companies Constantly Adjusting," *Computing Canada,* February 19, 1999.

MICHAEL FURDYK "Instead of focusing on how we can extract every single cent of revenue from the site, we focused on actually creating something that people would find useful. In doing so, we built a much more user-friendly product that was more focused on the experience for the person who was visiting it, instead of just on the bottom line."

GEN NEXT

INNOVATOR_ MICHAEL FURDYK, SERIAL ENTREPRENEUR, COACH INNOVATION_ FOUNDED MYDESKTOP.COM AT THE AGE OF 13 AND SOLD IT FOR OVER $1 MILLION TO INTERNET.COM BEFORE HIS SIXTEENTH BIRTHDAY

Most 13-year-old kids think entrepreneurial experience means setting up a roadside lemonade stand or delivering newspapers. At 13, Mike Furdyk did something a little bit different. He built an Internet enterprise. Now, at age 19, Furdyk is a seasoned entrepreneur and working hard at growing his third business. And when he's not incubating the next great startup, you might find Furdyk travelling around Canada to share his entrepreneurial insights, or in Redmond, Washington, telling the world's largest software company how to run its business. Microsoft hired Furdyk to give them a head start on how to reach the next generation of knowledge workers and consumers. He advises them on upcoming trends and product innovations, as well as on youth opinions and perspectives.

Clearly, Furdyk is not your typical teenager. He used his first six-month contract with Microsoft as a way of finishing his high school diploma. Fifteen credits short, he persuaded his Toronto principal to recognise his consulting work in lieu of going to class. Although at the time he preferred the Microsoft stint, Furdyk admits he hated to miss gym class but it was the only time he had to "stop working and just hang with friends."

The Internet enterprise he launched when he was 13 was a website called "Internet Exposed." Launched in 1995, in the early days of the Web, Furdyk's site helped its two or three thousand visitors per month understand this unfamiliar invention, the Internet. The budding entreprenur struck up an online friendship with an Australian counterpart, Michael Hayman, and the two tech-teen-wunderkinds developed an innovative online technology community called MyDesktop.com.

The new site quickly found an audience. Soon Furdyk went from being a kid who liked computers to being a kid who was getting big cheques; he began to rake in as much as $20,000 per month in advertising revenue. By 1999, he and his partners (including 20-something successful tech entrepreneur Albert Lai) were able to sell MyDesktop.com for over $1 million to a U.S.-based Internet.com, and Canada's *Profit* magazine named Furdyk one of the top 10 entrepreneurs who had "shaped the year."

Not content to rest on their successes, Furdyk and Hayman went on to create BuyBuddy.com. The new company is a provider of comparison-shopping infrastructure to Web portals and wireless carriers, and has already closed $4.5 million of second-round funding from leading Canadian venture capitalists XDL Intervest. But, for someone who made his first million before he was old enough to drive, Furdyk has never named moneymaking as a prime motivator. The mature teen and self-proclaimed "Conjuror of Wireless and Web Innovation" says he prefers to place an emphasis on finding value for his customers. "Instead of focusing on how we can extract every single cent of revenue from the site, we focused on actually creating something that people would find useful. In doing so, we built a much more user-friendly product that was more focused on the experience for the person who was visiting it, instead of just on the bottom line." The partners recognised that "online consumers required services and technology that would help them navigate the vast reaches of the Internet and choose intelligently from the infinite number of merchants and products available." Furdyk thinks of his latest corporate site as a prototype of "community-enabled commerce" and a "pioneer of B2B2C commerce." With this latest endeavour, aimed at empowering partners to create sophisticated, knowledgeable consumers, he hopes to lead the way in encouraging future entrepreneurs to develop responsible businesses.

Heralded as a role model for the youth movement by *FastCompany* magazine, Furdyk takes his real-life responsibilities very seriously and has set the pace for the next generation of global leaders. He now travels the world to cultivate the next generation of talent in the technology sector. Acting as an ambassador, he was recently invited by the Canadian government to tell the tale of his entrepreneurial success to every high school and college student in the Yukon. Furdyk welcomed the government's educational agenda and agreed to motivate rural youth by "getting them engaged and interested in using technology, especially because many are so isolated from traditional kinds of business." Articulate and confidant beyond his years, Furdyk is calmly championing the youth perspective and in doing so has become a role model for the next generation of ingenuity.

The commitment to make change by inspiring a community of young people was also the impetus for Furdyk's latest philanthropic project. Along with partner Jennifer Corriero (another rising star in the Canadian technology youth movement), Furdyk has founded *TakingITglobal.org*, a nonprofit organization that aims to be a virtual community centre for young people to "hang out" and share experiences and insights that matter to them—especially around the potential of technology. Furdyk and Corriero say the site is designed to give youth a voice and a creative place where they can discover their talent as leaders, entrepreneurs, or artists.

Firm in their realization that technology has the power to transform, Furdyk and Corriero built the site in part as an answer to their own questions: "What would we do if we could do anything? What if young people had access to meaningful opportunities that relate to their interests?" The resulting Web-based effort to promote positive change is now being funded by two Fortune 500 companies and the Royal Bank of Canada, and boasts members from 150 countries. Furdyk says he hopes the site, which includes a global youth events and scholarship database, an art gallery, and discussion groups, will "inspire people to be innovators of technology and to promote dialogue and debate of future technology issues."

Although he's already proven he's a staunch advocate for the power of youth, Furdyk admits that his age can still get in the way of success. Furdyk jokes that he still has to take cab rides when he's on business trips because he's too young to rent a car. When he decided to teach for the Toronto District School Board, he was forced to undertake a rigorous examination and interview process, because he was younger than most of his students. However, the rising entrepreneur, activist, and educator has proven that he can take the limitations of his youth in stride. With his considerable and noteworthy early success, Furdyk is one teen who has made good on Canada's promise to deliver the next generation of top technology talent.

TRUE CONTROL
IS BEING NEEDED

INNOVATOR_ DAN GELBART, SERIAL ENTREPRENEUR INNOVATION_ ESTABLISHED THE WORLD'S LARGEST INDEPENDENT SUPPLIER OF PRE-PRESS EQUIPMENT, LEADING THE DIGITAL TRANSFORMATION OF THE GRAPHIC ARTS INDUSTRY

In 1973, Dan Gelbart was a young Israeli engineer educated at the Technion, Israel's top engineering school. Looking for a warm quiet place to live and work ("for life," he says), he chose the Vancouver area—his search criteria were "an isotherm of warm climates and no wars." Even today, his loyalty to Burnaby and British Columbia is strong as ever. The company he founded in 1983, Creo, now has sales approaching $1 billion (Can.) per year and over 4,000 employees. Almost 20 years after its creation, it is still one of the fastest-growing companies in the country. And it is still run out of Burnaby.

Creo is one of BC's greatest high-tech success stories. One of the largest technology companies in the province, it exports almost 100 percent of its product. And its innovative approach to doing business has also been recognised. Creo has been named one of the country's top 50 ethical stocks and one of the 100 best companies to work for.

Its revenues have surpassed even those of the company that launched Gelbart into entrepreneurship—the long-established firm of MacDonald Dettwiler and Associates. Recognising that this young R&D engineer was more than a straight researcher, MDA spun off two high-tech startups, both successful, based on Gelbart's patents.

While at MDA he had met Ken Spencer, a seasoned manager, and together they started "looking for a big business that would grow," Gelbart says. In 1983, they set out to build an optical tape recorder, a device that could store vast amounts of data for such applications as maps, films, or satellite downloads. They took on some of the top electronics companies in the world, and won. Creo became a leader in electronics research and in high-precision manufacturing. As Gelbart proudly notes, the products were "good enough to be exported to Japan." Ken Spencer retired, but Gelbart continued to run the successful entreprise.

The real growth for Creo, however, came when Gelbart brought in Amos Michelson, who had an established reputation in Silicon Valley. "We started out in data storage, but that did not grow fast enough for our tastes, so we sold it. We thought about semi-conductors, but they're very cyclical. That's when we came up with printing," Gelbart says. On reflection, he feels that his initial focus on building the best product, while successful, was misguided. Michelson's vision was not to be an OEM but to be a complete systems supplier. He looked to the market to find the best place to use their research and product development skills.

Their core R&D and manufacturing competency was a mix of optics, electronics, software, and hardware. Their chosen market was printing, but a specific segment of the printing market: the computer-to-plate, or CTP, process for the printing industry, to make the entire pre-press phase of printing both electronic and digital. For the next few years, most of the company's R&D was devoted to building a viable thermal imaging engine for CTP applications. In 1994, Creo delivered its first commercial system.

Michelson forged an alliance with printing industry giant Heidelberg AG, the largest printing press manufacturer in the world, gaining worldwide distribution of Creo's products. Today their products are the industry standard. Creo is the world leader, with three times the market share of its next competitor. The company has also transformed itself, over the last 10 years from an engineering-based product company into a solutions provider.

CREO PLACES A GREAT DEAL OF EMPHASIS ON STAYING CLOSE TO ITS CUSTOMERS. EVERY YEAR IT HOSTS A USERS CONFERENCE TO HELP ENSURE CREO CUSTOMERS HAVE THE OPPORTUNITY TO SHARE INFORMATION AND EXPERIENCE. IT HELPS ENSURE THAT THEY GET THE MOST OUT OF THE COMPANY'S PRODUCTS AND PLAY A ROLE IN SHAPING THE COMPANY'S FUTURE DIRECTIONS. CUSTOMERS ARE INCLUDED THROUGHOUT THE PRODUCT DEVELOPMENT PROCESS, ESPECIALLY THE EARLY PILOTS OF NEW PRODUCTS. ANALYSTS HAVE NOTED THAT "WHILE CREO'S ISN'T THE CHEAPEST PRODUCT ON THE MARKET, IT'S ARGUABLY THE BEST."

Creo places a great deal of emphasis on staying close to its customers. Every year it hosts a users conference to help ensure Creo customers have the opportunity to share information and experience. It helps ensure that they get the most out of the company's products and play a role in shaping the company's future directions. Customers are included throughout the product development process, especially the early pilots of new products. Analysts have noted that "while Creo's isn't the cheapest product on the market, it's arguably the best."

Innovation is central to Creo's culture. The name Creo is Latin for "to create," and the company's slogan is "Imagine, create, believe." With over 300 patents awarded or pending, it devotes more than 10 percent of its revenue to R&D. But to achieve its growth and market presence it has also made a number of significant acquisitions and partnerships. In 2000 it acquired its major (and much larger) Israeli-based competitor, Scitex Corporation. The Scitex acquisition was motivated in part by a desire to broaden the product range, strengthen technology support, and bulk up in R&D. In 2002 Creo made a major investment in printCafe Inc, a specialist in print management systems software, to develop a business-to-business communication solution tailored for the graphic arts industry. At its most recent user conference Creo piloted Internet-based products and is focusing on moving into that space more aggressively.

Gelbart has no regrets at all about choosing the Vancouver area as the place to live and work. He firmly believes that good R&D can take place anywhere in the world if the social environment is right. His researchers love B.C. There are good universities, a local high-tech culture—and it is without question one of the most beautiful cities in the world. As a result, Creo has very low employee turnover. Gelbart does admit that running a global business from Burnaby is a bit challenging, but says that these challenges can be managed and, on balance, it is well worth it.

Unlike many entrepreneurs, Gelbart has kind words for the government, both at the federal and provincial levels, and he credits the government with the ability to see and support Creo's vision. While Creo was initially financed from Gelbart's and Spencer's own funds and by doing some contracted R&D work, the B.C. Science Council and the National Research Council also provided early funding. The Government of Canada was one of the first customers, with the world's first functional optical data recorders being used by the Ministry of Energy, Mines and Resources. Venture capital became important later in Creo's growth, leading to a successful IPO in 1999.

Before Gelbart started his own company he believed that timing and luck played a big part in entrepreneurial success. Now he argues that executing a business plan well is more of a science.

Gelbart would be the first to say that being an entrepreneur does not mean that you should be the CEO. He also argues that real control comes from being needed by the corporation, not from how many shares you own. An important key to success is knowing your own strengths and weaknesses. He sees himself as a practical person and a researcher who "did research in order to make a product to make money." Although he and his partner, Ken Spencer, did a reasonable job of taking Creo though its startup phase and early success, he credits Creo's rapid growth in the '90s to the decision to hire Michelson, and his "second-generation" team of managers. Handing over the management of the company has brought in the experience and skills needed to succeed in the big leagues and has allowed Gelbart to return to his favourite role—that of chief technology officer.

He emphasises the importance of building a team and sharing the wealth, but it is much more than just lip service. Every CREO employee is a shareholder. In fact, staff owned half of the company's equity prior to its IPO. The company's approach to performance appraisal is also unique—every year employees are given a performance review not just by their immediate supervisors but by every one of their colleagues. "It eliminates insecure bosses who don't want anyone smarter coming along and taking over their jobs," Gelbart says.[1] And the investment in employees has paid off. Staff turnover has always been low—in fact, turnover was zero in the company's first seven years. And the first person to leave came back a year later.

While Gelbart could never imagine becoming a venture capitalist, he recognises that he now thinks much more like one, with an increased focus on share price. While encouraging young entrepreneurs to focus on their own ideas and dreams, he cautions them, "Venture capitalists think the way you will in 20 years."

Today, Gelbart is still doing what he loves—research. He continues to look for the practical applications and often still works from his lab in his own home—just as he did when he launched Creo. The main difference is that the home is a bit larger now.

[1] "Technology Entrepreneurs—Dan Gelbart and Amos Michelson," *B.C. Business Magazine,*

LESSONS MY WIFE TAUGHT ME ABOUT TECHNOLOGY

INNOVATOR_ JAMES GOSLING, INFRASTRUCTURE BUILDER, ACADEMIC BRIDGE INNOVATION_ CREATOR OF JAVA, ONE OF THE WORLD'S MOST POPULAR COMPUTING LANGUAGES

Today when you say Java most people think programming language, not coffee. But few know that one of the world's most popular computing languages was developed by a Canadian, James Gosling, born and bred in Calgary, Alberta. A vice president and fellow at Sun Microsystems, he describes himself as "a pretty serious geek." By the time he was 14, Gosling had taught himself enough about programming to get a part-time job writing software at the University of Calgary. He showed glimmers of technical genius at a very early age. While studying for his undergraduate degree in computer science at the University of Calgary he developed a text editor called "Emacs," which became the most widely used UNIX text editor in the world.

But by his own account Gosling never set out to change the world and was not particularly ambitious. "I never gave a moment's thought to the word 'career.' I just liked to play with cool toys." And he has played with and built his fair share including satellite data acquisition systems, mail systems, compilers, and text editors. But he is best known for Java. In 1997, *PC Magazine* awarded Gosling and his team its "Person of the Year" award for technical excellence—"It's hard to imagine a product that has had more influence on the way we think about personal computing than Java." The Canadian Productivity Awards (CIPA), which added Gosling to the organization's Hall of Fame in 2001, noted, "Studies have shown that the productivity of software developers using Java is two to five times greater than with any other programming language." Its effect on programmer productivity may go a long way toward explaining its popularity as well as its impact. Today Java is turning up everywhere from mainframes to mobile phones to games to "smart" appliances.

Curiosity and technical genius aside, what lessons do Gosling's experience offer for other innovators? Despite his passion for "building neat stuff," Gosling insists that successful innovations are driven by the market and by customers, not by technology. In the early nineties, Gosling and some colleagues at Sun were working on the "Green Project," "trying to figure out what the 'next wave' of computing would be and how we might catch it. We quickly came to the conclusion that at least one of the waves was going to be the convergence of digitally controlled consumer devices and computers." Before long, the team developed a handheld wireless PDA, and was working on a programming language that would allow multiple types of consumer electronic devices to communicate with each other. It was this language that became Java.

From the beginning, the Green Project team understood that business models and end users were as important as technology. As Gosling says, "New technologies are neat. I'm one of the worst for getting excited about new technologies. But if you've floated a few neat technologies around you quickly discover that people say, 'So why should I be interested?' One person's perception of neat is not necessarily another's. And most people don't buy something because it's cool. They buy it because it helps them—it gets a job done, it has some practical value, it improves their life in some way."

Knowing that a new product needs to offer a solution to a problem doesn't make it any easier to develop a winning innovation. "In some sense you have to start from a germ of a technological idea, because it's hard to reason about markets or products in a vacuum. You have to think about both of them at the same time, but you have to let the market lead. The technology is there to solve problems, and if there isn't a problem to solve, then it's a curiosity, not a solution."

TAKING BORCZ'S ADVICE, GOSLING WROTE A SHORT PAPER THAT OUTLINED THE BENEFITS OF JAVA, AND WHY THEY WERE IMPORTANT FOR DEVELOPING SOFTWARE. RATHER THAN PITCHING JAVA IN TERMS OF "HERE ARE ALL THE COOL THINGS IT DOES," THE PAPER PITCHED IT BY SAYING, "HERE ARE ALL THESE HARD PROBLEMS THAT YOU HAVE THAT IT SOLVES." IN DOING THIS, GOSLING MANAGED "TO EXPLAIN THINGS TO PEOPLE IN TERMS OF THE THINGS THAT WOULD APPEAL TO THEM RATHER THAN THINGS THAT WOULD APPEAL TO ME."

So was Java a curiosity, or a solution? Was its development being led by the market? Did the market know what Java was for when it was introduced, and could an avowed techie sell the concept? Ask Gosling these questions, and he'll explain what he learned from his wife.

"Java had been used inside Sun for about four years by a small, enthusiastic crew. I was looking forward to getting it out there so that others could enjoy it too. I tended to think of it in pretty technical terms. This was a fairly serious problem since relatively few people could make sense of what I was babbling. My wife, Judy Borcz, is a very bright person, but not a geek (she's a Wharton MBA type). So when I would erupt in fits of techno-babble, it didn't help her understanding at all. She kept saying, 'You're so excited about this thing; why should *I* be excited about it? Explain it to *me*.' She pushed me very hard to write down, in terms that mattered to people like her, what was so cool about Java."

Taking Borcz's advice, Gosling wrote a short paper that outlined the benefits of Java, and why they were important for developing software. Rather than pitching Java in terms of "Here are all the cool things it does," the paper pitched it by saying, "Here are all these hard problems that you have that it solves." In doing this, Gosling managed "to explain things to people in terms of the things that would appeal to them rather than things that would appeal to me."

> **THE TECHNOLOGY IS THERE TO SOLVE PROBLEMS, AND IF THERE ISN'T A PROBLEM TO SOLVE, THEN IT'S A CURIOSITY, NOT A SOLUTION."**

As Gosling suggests, the ability to understand what appeals to "them" (the consumers) rather than "me" (the producer) is crucial, yet it is often difficult to achieve. One of the reasons this is such a challenge is that many potential benefits for consumers are not easily identified in the early stages of innovation. But if consumers can be enticed to try a new product or service, they may discover benefits on their own that weren't articulated by the designers.

A recent *CIO* magazine feature describes Java's evolution since it was released. "Java's ascendancy hasn't happened quite the way Sun envisioned back in 1996. In stark contrast to the swarm of Java applets populating the Web during its first years, client-side Java is almost nonexistent today. Instead, the language has moved behind the scenes, within the application servers that drive corporate websites—and, increasingly, companies' line-of-business applications. During the past year, enterprises have taken Java to heart like never before. The language has matured. Tools for developing and deploying heavyweight

Java applications are readily available from Borland, IBM, and Sun. And developers now have a wealth of experience with the language."

The experience with Java shows that customers often find uses for products and technologies that were unanticipated by their developers and, interestingly, developers may have to adapt their strategies to keep up with their customers. Not all new products and technologies will experience such a shift in how they are adopted, but developers and businesses must be prepared to grow alongside their innovations, providing the necessary support to enter new markets and to serve new customer groups. When customers do lead the way, smart businesses will figure out how to follow. So Gosling's team continues to enhance Java, exploring its uses in real-time systems like industrial robots, as well as "smart" cars, airplanes, and appliances.

Gosling is recognised as a titan in the programming world. Although his achievements have had worldwide impact, and he has lived in the United States for 22 years, he is still proud of his Canadian roots. "I go back to Canada regularly," he says. "All my family is there. I certainly feel culturally Canadian. The Canada I grew up in had a much more laid-back view of the universe." He also continues to support his alma mater, directing prestigious awards to the university's scholarship funds. And Gosling's Web page reveals he has never lost his Canadian modesty or sense of humour. Visitors logging on to www.java.sun.com/people/jag. are greeted with "Hi, I'm James Gosling, and this is my home page. I'm a guy that works at JavaSoft, a division of Sun Microsystems, where I do odd jobs like helping out with the system architecture and wandering around the country giving talks like why Java is the greatest thing since sliced bread." The jury is out on whether Java is the greatest thing since sliced bread, but it most certainly is the application that changed the computing world forever.

WEDNESDAY NIGHTS

INNOVATOR_ RUBIN GRUBER, SERIAL ENTREPRENEUR INNOVATION_ FOUNDED SEVERAL WORLD-CLASS STARTUPS, FROM CALL CENTRE SYSTEMS TO MULTIMEDIA SERVERS TO DEVELOPING THE NEW FACE OF THE PUBLIC NETWORK

Rubin Gruber is a serial entrepreneur whose focus on vision and teambuilding have produced remarkable results. Sonus Networks, his fifth company, which he founded in 1997, broke $100 million in sales by 2001 and was named one of America's most dynamic companies by *Forbes* magazine.

Perhaps most remarkable about Sonus, and a testament to Gruber's unique vision and style, is the fact that during the first four years of operation, there was no employee turnover. None. Not a single employee quit the startup company, a fact Gruber is extremely proud of. Management's success in retaining employees at Sonus Networks (a provider of voice infrastructure products for public networks) is a testament to the strong culture as well as the company commitment to developing a place where people "work hard and play harder."

Gruber, who grew up in Montreal and did his first degree at McGill, knows from his own experience that working in a startup company can be extremely stressful. As Gruber says, "No one is going to tell you to work 10 to 15 hours a day," but people know what they have to do, and it often does involve long hours. "You have to commit to doing it," notes Gruber, but he recognises that this commitment involves an employee's entire family. So at Gruber's companies, families are encouraged to come to the workplace. "You can't give up on your family. You can bring your family here." And in the early days that's exactly what Gruber's employees did. Except on Friday nights, when Gruber suggested everyone go home and spend quality time with the family there.

BUT SINCE BUSINESS PATTERNS ARE CYCLICAL, ONCE YOU IDENTIFY WHERE YOU ARE IN A CYCLE, "YOU ALMOST KNOW WHAT'S GOING TO HAPPEN." CREATING NEW BUSINESSES IS ALL ABOUT RECOGNISING PATTERNS AND EVENTS, SO THAT YOU KNOW WHAT WILL HAPPEN NEXT. "IF YOU CAN RECOGNISE THE SIZE AND TIMING OF OPPORTUNITY, YOU CAN START A COMPANY THAT WILL BE SUCCESSFUL."

This way of operating was already second nature for the first 100 employees at Sonus Networks. That's because many of them had previously worked with Gruber or Mike Hluchyj, Sonus's co-founder, at other companies. Finding exceptional employees who will "fit" with the Sonus vision and culture is no mean feat. The screening process is often as unusual as it is rigorous. The company didn't use headhunters to find employees, and it didn't hire any trainees. But being a known quantity was no guarantee of being hired by Sonus. Gruber says that an interview with him is often "the worst interview" in people's careers. He doesn't ask "the questions in the book," so there's really no easy way for potential employees to prepare. His questions can be "totally obtuse," and satisfying just Gruber is not the challenge. Typically, a candidate is interviewed by up to 10 people. At some point Gruber will turn the questioning over to the interviewee, but not for the standard questions about the company. He expects his potential employees to ask him intriguing questions about himself, and if he doesn't like a question, he doesn't hesitate to tell the person that the question was a dud.

Employees quickly learn that there's no place for ego in a Gruber-run company. Everyone is an equal player on the team, and there are no superstars. Even when Sonus Networks went public several years ago and many of the engineers working at the company became millionaires, they didn't leave—because they like working there.

The environment may be intense and demanding, but Gruber is also preoccupied with keeping it healthy and productive. He learned that he didn't want to deal with office politics, but wanted to rise and fall on his own merits. (In fact it was a "political" incident at a company he worked for that was the catalyst for his decision to become an entrepreneur.) He simply extrapolated from his own experience—"As soon as the politics appear, it's not fun anymore"—to build an attractive working environment. He believes that you can avoid politics when everyone is committed to a shared vision and sense of purpose for the company. "You hire the right people, and it falls into place. I rarely have to tell employees what to do. The right people know their part and do it." Although he doesn't tell employees how to do their jobs, Gruber and his management team invest a lot of effort in articulating and reinforcing the corporate vision. "We'll tell you exactly *what* you have to do, and describe success so you'll know when you get there. We repeat the vision every step along the way." Gruber says his companies are not democracies. The vision is driven from the top of the company. As he says, "It takes every ounce of management talent to spend time to do it; it's not easy."

Management gurus write about the importance of corporate visions, but Gruber's commitment to vision is not just a passing fancy. He believes that when companies stop reinforcing their corporate visions, they make mistakes. So one thing that Gruber has been doing, for close to 30 years, and with all his startup companies, is the Wednesday night meeting. If you work with Gruber, Wednesday night belongs to the company, as he puts it. That's when the entire company meets with the management team. The meeting keeps lines of communication open, and allows anyone to ask any questions about the corporate vision and direction. By making the vision transparent, on a weekly basis, the worst thing that can happen is that someone misinterprets the vision for the six days between meetings.

> GRUBER SIMPLY EXTRAPOLATED FROM HIS OWN EXPERIENCE—"AS SOON AS THE POLITICS APPEAR, IT'S NOT FUN ANYMORE"—TO BUILD AN ATTRACTIVE WORKING ENVIRONMENT. HE BELIEVES THAT YOU CAN AVOID POLITICS WHEN EVERYONE IS COMMITTED TO A SHARED VISION AND SENSE OF PURPOSE FOR THE COMPANY. "YOU HIRE THE RIGHT PEOPLE, AND IT FALLS INTO PLACE. I RARELY HAVE TO TELL EMPLOYEES WHAT TO DO. THE RIGHT PEOPLE KNOW THEIR PART AND DO IT."

When Gruber first started working, at the General Motors Research Laboratories in Warren, Michigan, his boss showed him a big machine in a glassed-in room, and told him he'd do some work on it. It was a computer, something Gruber had never heard of before. Within a few years he not only knew much of what there was to know about computers but had founded Cambridge Telecommunications (CTX), the first company to incorporate microprocessors in communications equipment and an early supplier of packet network access equipment. (CTX was acquired by GTE and remains part of Sprint.) He also served as senior vice president at BBN Communications Corporation, where he introduced a new generation of packet-switching equipment in support of mission-critical applications. Gruber then founded Davox, the leading supplier of outbound call-centre systems. And just prior to Sonus, he founded VideoServer, Inc. (now ezenia!), a leading provider of multimedia communications servers.

For more than 30 years, Gruber has built many businesses around computer technology, and he's not ready to quit just yet. He actually retired in 1996, but within a year he was pitching the idea for Sonus Networks.

How does Gruber develop new business ideas? Asked this question, he starts by saying that he always needs a six-month break between businesses. Reflection is important, he says: "You learn a lot every time, and you learn from your past mistakes." The break also provides a much-needed perspective on the trends in the market. When he's involved with a business, he doesn't have the perspective he needs to understand where new business opportunities lie. But since business patterns are cyclical, once you identify where you are in a cycle, "you almost know what's going to happen." Creating new businesses is all about recognising patterns and events, so that you know what will happen next. "If you can recognise the size and timing of opportunity, you can start a company that will be successful."

For Gruber, being a successful entrepreneur involves identifying a market opportunity, creating and nurturing a corporate vision, and finding the right employees to work with. When he's assessing investment opportunities, he always looks at the people first. Passion and leadership are key. Gruber would rather invest in "a passionate 'A' player with a 'C' plan, than a 'B' player with an 'A' plan." Good people can fix weak plans, but if Gruber doesn't trust the person looking for support, he won't invest in the business. He's also wary of people who start businesses just for the money, those with a great idea but lots of competition, and those who choose to move away from their families to develop their businesses. Putting people first has proved to be an important key to his success.

Hamnett Hill Austin Hill_ Hammie Hill_

MISSIONARIES AND MERCENARIES

INNOVATOR_ AUSTIN, HAMNETT, AND HAMMIE HILL, SERIAL ENTREPRENEURS INNOVATION_ DEVELOPERS OF "THOUGHT LEADERSHIP" INTERNET PRIVACY SOFTWARE

Zero Knowledge has been called a magnificent obsession. The company was built on Austin Hill's evangelical fervour for protecting privacy, his brother Hamnett's operational know-how, and the financial wisdom of their father, chartered accountant Hammie Hill. Already successful entrepreneurs in their teens, the Hill boys and their dad founded the company on the insight that growing concerns over Internet privacy represented a large market opportunity. The company evolved from providing products aimed at individuals to providing integrated solutions for large corporations. The founders attribute the success of their company to balance. They have combined a missionary zeal for their corporate vision with a hard-nosed "mercenary" approach to developing the business.

As Hammie notes, "Often you perceive a problem, but there's no defined market that you can analyze and say, 'Spending to protect privacy on the Internet is X million or billion a year.' It was early in the whole evolution of the Internet, and even the definitions of privacy issues were unclear. You often end up going on gut feel, knowing that there is a big problem here that people will spend money to solve." Austin adds, "We saw a general trend and we took a risk, saying, 'Well, let's just take a stab at something.'"

Achieving balance between vision and adaptability is crucial, as Austin explains. "In building a good company, you have to have enough confidence and faith so that you're not getting shut down. At the same time, you must be smart enough to take every single piece of criticism, and understand every single reason why something is not going to work, and plan for a way around it—innovate around it." While innovating around challenges, the company's vision has evolved. "The first focus of the company was creating absolute privacy on the Internet, which presented a huge technical challenge. The total focus was really on solving technical issues and building a product that worked. Initially we very much attached ourselves to the advocacy cause of privacy, not only because we personally believed in it, but also because that was our market. As we learned more about the market and what consumers wanted and what our partners wanted, we evolved. We were able to put the focus on the business of being a solutions provider."

A venture capitalist told Austin that there are two types of entrepreneurs: missionaries and mercenaries. "Missionaries have faith, vision, a dream. And they tend to be very passionate about it. Mercenaries tend to come at the starting of a business and say, "'There is $100 million spent in this market; I'm going to go in and grab 20 percent.'" The key is to combine missionary zeal with "mercenary" business sense. "From day one we said, 'If we can't make this pay, if this just ends up being an interesting technology or a strong advocacy position, then we're not doing our job.' We have to figure out how to make a business." Hammie comments that "a pitfall a lot of entrepreneurs can face is the idea that starting a business is just a way to earn an income. The vision must be much bigger than that. From the start we wanted to create a company that we could ultimately either take public or sell for a significant capital appreciation. Second, we ran it, from the beginning, as if it were a public company. So the decisions are made on rational analysis and not to specific family interests."

The synergies in the family are obvious. Hammie has the background in finance and raising money. Austin is the evangelist, the ideas guy. Hamnett's skill is operational—he can build a team around a goal to make things happen. The family works together like a well-oiled machine, even finishing each other's sentences. Says Austin, "We were a team from the beginning. We didn't have to spend time developing trust; it was there from the start. One of the most important things is the ability to pick the team that you're going

---

THE FIRST FOCUS OF THE COMPANY WAS CREATING ABSOLUTE PRIVACY ON THE INTERNET, WHICH PRESENTED A HUGE TECHNICAL CHALLENGE. THE TOTAL FOCUS WAS REALLY ON SOLVING TECHNICAL ISSUES AND BUILDING A PRODUCT THAT WORKED.

---

to build the company with right across the board: what are the values that you're looking for in your team members? And you have to be able to say, 'In two years, what do I imagine this company being?' Not only in terms of the product and the business execution and the money, but the type of people—the culture."

Hammie was always involved in a number of businesses on the side: real estate ventures, oil ventures, and some software companies. Austin notes, "That excitement about a new idea, or a new company, would be something that was really shared by the family. It would be the type of thing where, 'If this hits big, we may all go to Disneyland.' When a business didn't go well, we discussed it. So we had this natural sense of some of the ebb and flow of business." Hamnett agrees the entrepreneurial family culture played a role. "I remember dinner conversations about the Alberta Stock Exchange, and this oil player, that oil player." Austin remembers playing the stock market when he was 10 years old. He made $900 and bought a great stereo. "So I had an appreciation of the idea that investing in good ideas, at the right times, was an exciting thing."

ONE OF THE MOST IMPORTANT THINGS IS THE ABILITY TO PICK THE TEAM THAT YOU'RE GOING TO BUILD THE COMPANY WITH RIGHT ACROSS THE BOARD: WHAT ARE THE VALUES THAT YOU'RE LOOKING FOR IN YOUR TEAM MEMBERS? AND YOU HAVE TO BE ABLE TO SAY, 'IN TWO YEARS, WHAT DO I IMAGINE THIS COMPANY BEING?' NOT ONLY IN TERMS OF THE PRODUCT AND THE BUSINESS EXECUTION AND THE MONEY, BUT THE TYPE OF PEOPLE—THE CULTURE.

Hamnett emphasizes, "One of the fundamental factors in being an entrepreneur is having massive risk tolerance. Growing up seeing some of the ups and downs, financially, that we went through as a family, we knew that the world doesn't end if things are tight for a while. So you get used to that and it's not so scary to think that you don't have a job, because you're confident you can always make money. The world is not going to come to an end if I'm broke for a little while. Having that risk tolerance allows you to go out there and take a good chunk of your net savings and throw it at this wacky privacy idea, or this wacky Internet idea."

Although their market is global, Canada was an obvious place to stay to do business. "Canada had a fairly good security and software industry, because it always looked at the market a bit more favourably than the U.S. There were some regulatory issues that made Canada attractive to us. We also needed talent and had to be able to recruit the best. Often what sold new recruits was just quality of life. For younger people, Montreal is a very cost-effective city, and it's also a big multicultural city. When our developers compare that to going to San Francisco, they might have had a salary that was 40 percent more, but they'd be living with three roommates and sharing carpools." Hammie adds, "When we've brought guys out of California up here and some of them have actually moved up with their families, they're amazed they can walk down the street with their wives, at 10 o'clock or 11 o'clock at night, and not even have a second thought about their safety. Canada is a safer place to live. The business climate is not as brutal as it is in the States. There's a bit more openness, a bit more understanding of people and that the rules of the game are slightly different. The competitive climate among your employees is not as cut-throat as some areas in the States."

It is important, however, for Canada to continue to build its entrepreneurial culture. "I think that Canada has done a lot better in keeping some of its two- and three-time entrepreneurs and keeping some of its top business executives. We have started to see the creation of an angel environment, which is critical to support entrepreneurs. The angel and two- and three-time entrepreneur community in Canada is actually quite closely knit. We all tend to know each other, and we'll refer new entrepreneurs to each other. That type of network, or ecology, of mentors, angels, entrepreneurs, and connections, is critical to building a continual engine of innovation. And Canada is only beginning to see that foundation start to be laid down. A lot of the changes that the government has made, in terms of taxes, capital gains, have been favourable to that. We need to ensure our successful entrepreneurs stay."

THE GENOMICS
OF PROFITABILITY

INNOVATOR_ DR. JULIA LEVY, SERIAL ENTREPRENEUR, ACADEMIC BRIDGE, COACH
INNOVATION_ ONE OF ONLY A HANDFUL OF ENTREPRENEURS IN THE WORLD TO TAKE A
BIOTECH COMPANY FROM STARTUP TO PROFITABILITY

"If anyone had told me in 1981 that I would be the CEO of a major biotech company in Canada by 2000, I would have thought they were certifiably crazy." That's how Levy describes her entry into the world of biotechnology entrepreneurship.

Levy co-founded QLT Inc in Vancouver in 1981. Nineteen years later, it posted its first profit. For most businesses, that would be pretty poor performance, but not for QLT. Research and development in biotech is a long-distance event, not a sprint, and there are many barriers to success. In fact, QLT is only the second Canadian company to complete the process, and, according to Levy, one of only about 14 out of some 400 worldwide to go from startup to product success and profitability.

Back in 1980, Levy was a successful researcher at the University of British Columbia. She had come to Canada in 1940 as a young refugee from Singapore. She pursued a research degree in immunology, studying at UBC and at the University of London, where she gained her Ph.D. in immunology.

She admits to having been very happy as a university researcher and was prominent in her field, a full professor at UBC with more than 20 years of experience, when she was asked to join three of her colleagues and co-found a biotech research company—then Quadra Logic Technologies Inc. Her research interests had led her to consider the opportunities for light-activated drugs, inspired by burn-like rashes her young son had acquired in the fields in B.C. This pioneering field of photodynamic therapy combines the administration of an intravenous drug with a low-level non-thermal laser light trigger. It takes both elements to produce an effect.

> SHE SUGGESTS THAT IN RESEARCH-DRIVEN COMPANIES, THE SAME BLEND OF ACADEMIC ARROGANCE AND KNOWLEDGE OF HOW SMART YOU ARE THAT CAN LEAD TO DISTINCTIVE RESEARCH CAN ALSO LEAD TO AN UNREALISTIC BELIEF IN YOUR CAPABILITY TO BRING SOMETHING TO MARKET, WITH POTENTIALLY DISASTROUS CONSEQUENCES.

QLT's early focus was on the development of a cancer treatment called Photofrin. This effort was the impetus that brought Levy out of the lab and into the wider business world and was the company's first success. As a key scientist with QLT she was still very much working in a research environment. She held an industrial professorship, which allowed her to hold a senior position within QLT while continuing as an academic at UBC. In this way, the basic research was carried out at UBC, while QLT focused on the product development. This creative university/industry partnership was complemented by early research grants from Canada's Natural Sciences and Engineering Council (NSERC).

QLT's founders had always planned to focus on the R&D side of drug development and had expected that, as they moved through the stages from early development to clinical trial and manufacture, they would partner with more experienced and larger pharmaceutical companies to bring the drug to market. Early experiences with Johnson & Johnson and American Cyanamid, while useful, did not turn out quite as planned; QLT ended up taking over control of the whole development process, thus gaining invaluable experience in end-to-end drug development.

But it was QLT's second drug that has had the real impact. The leading cause of blindness in people over the age of 50 is the "wet" form of a disease called age-related macular degeneration, or AMD. Every year some 500,000 people develop this form of the disease, which can lead to irreversible blindness in a period of months to a few years. Using their world-leading research capabilities in photodynamic therapy, QLT discovered a drug—Visudyne—that can treat AMD, and they used the early experience gained with Photofrin to take it to clinical trials. They then partnered with the CIBA Vision subsidiary of the Swiss firm Novartis AG in the production and marketing of this new drug. Along the way, they sold off the rights to Photofrin to focus on Visudyne and further efforts in photodynamic therapy.

QLT did not look to government for grants to set up their business (other than those early NSERC grants). Nor did venture capitalists look kindly on knowledge-based industries—especially in the resource-based B.C. economy of the '80s. So Julia and her colleagues looked initially for educated insiders—mainly from within UBC—for the first $500,000. Since then, they have looked to the markets for funds, initially going public on the VSE to raise about $3 million. Their partnership with (and a major investment from Cyanamid) helped them move to the TSE and raise additional funds.

QLT is now a mature biotech company with two successful drugs brought from the lab to the market. The company is now working toward a third drug—an amazing Canadian success story.

According to Levy, QLT's ranking as one of the top 50 employers in Canada is due to the way they recruit and retain workers. Her description sounds idyllic. "You want to make it a great place to work. It's a science-based company. There are a lot of egalitarians. We have our own gym, and a personal trainer who comes in twice a week for employees, and shiatsu, and they have their own bistro, and … it's a nice place to work! And, we pay well." To build a successful company, QLT's almost 400 people have been recruited from across the world, and within a few years they become strong and vocal Canadians.

Levy has received many awards as a researcher and as a business executive, including Fellow of the Royal Society of Canada, Pacific Canada Entrepreneur of the Year, and Canadian Women Entrepreneur of the Year (in 1999). She was appointed an Officer of the Order of Canada in 2001. But her success has a much more personal reward. Her mother lost her sight to AMD in the 1980s, and Levy feels that her research is her mother's legacy.

In looking back at QLT's success, Levy offers sound advice to startup companies: "Never over-promise 'the street.'" She sees this as a particular sin of biotech firms, but other observers might see it in other high-tech ventures as well. She suggests that in research-driven companies, it is the same blend of academic arrogance and knowledge of how smart you are that leads to distinctive research that can also lead to an unrealistic belief in your capability to bring something to market, with potentially disastrous consequences.

She also encourages people with ideas to follow them through, and not to listen to the naysayers, who will always tell you why it can't be done. Instead, she suggests seeking out the people who have the real knowledge and experience and listening carefully to their war stories and suggestions.

Levy stepped up to the CEO position in 1995, having been the company's chief scientist since the mid-1980s. While she claims to be an introvert and does not like to be the centre of attention, working as the chief scientist put her in front of potential investors, partners, and industry analysts. She discovered that she enjoyed learning all the aspects of running a drug development company and, after a couple of failed attempts to find the right CEO, agreed to step up and take it on. She suggests that women seldom aspire to become the CEO. Rather, in the drive to achieve other objectives (in her case, she wanted to change the world thorough medical research), they discover that they have the capability to be a leader and step up to the opportunity, often reluctantly. An undoubted success as QLT's CEO, she was a firm leader, but had a strong management team and listened carefully to what they had to say.

SHE ALSO ENCOURAGES PEOPLE WITH IDEAS TO FOLLOW THEM THROUGH, AND NOT TO LISTEN TO THE NAYSAYERS, WHO WILL ALWAYS TELL YOU WHY IT CAN'T BE DONE. INSTEAD, SHE SUGGESTS SEEKING OUT THE PEOPLE WHO HAVE THE REAL KNOWLEDGE AND EXPERIENCE AND LISTENING CAREFULLY TO THEIR WAR STORIES AND SUGGESTIONS.

Now in her mid-60s, Levy has no intention of retiring. While she has recently handed over the president and CEO position at QLT, she remains active as the executive chairman of its scientific advisory board. She also sits on the board of directors at QLT and a number of Canadian startup biotech companies, and continues to share her knowledge and experience.

DON MATTRICK "The thing that I've learned, and that we've learned as a company, is that there are amazing people inside Canada. The quality of the people, their ability to excel, both technically and creatively, is one of the reasons why our company has succeeded the way it has."

HE GOT GAME

INNOVATOR_ DON MATTRICK, SERIAL ENTREPRENEUR INNOVATION_ BUILT THE WORLD'S LARGEST INTERACTIVE ENTERTAINMENT SOFTWARE COMPANY

Like many of those who have made it big in Silicon Valley, Don Mattrick started in the business as a teenager looking to earn money for university. But he also just wanted to have fun. Twenty years later not much has changed, except perhaps the scale of his playground. He is still devoted to creating top-notch gaming experiences, but now he's at the helm of one of the world's most successful entertainment companies—Electronic Arts, the world's largest video game maker. As president of EA's Worldwide Studios, in Vancouver, he is still in touch with his inner child. Mattrick shifts seamlessly between enthusiastic discussions of the intricacies of his latest game and sophisticated analyses of the complex corporate strategies needed for growth and building market share.

Fortune magazine has suggested that the video game industry exists in a kind of parallel universe. Despite the recession, the tech slump, and September 11, sales of video game hardware, software, and accessories grew by over 40 percent in 2001 to a record $9.4 billion (U.S.). Leading the pack is EA, with about 12 percent of the market and assets valued at around $8.7 billion.

Mattrick remains matter-of-fact and remarkably unassuming about his phenomenal rise in fame, fun, and fortune. Growing up in Burnaby, B.C., he worked at a number of jobs in order to save money for university. After sweating though summers working in grocery stores and on construction sites, he wanted to find something that would be more fun. Although computers were still a novelty in the late 1970s, Mattrick decided that working in a computer store would fit the bill. When the manager of a local store told him they were not hiring, he offered to work for free. From the start, he showed an exceptional ability to interact with people and answer their questions. Perhaps the fact that he was only 15 at the time made him less threatening. The more he worked, the more he learned more about computers and how to program them. He also learned a lot about consumers' likes and dislikes. As he gained knowledge and confidence, he thought it would be fun to try to create a product of his own. His goal was "not a grand plan to build a company. It was just to see something I had created being sold at retail."

Working with another teenager—Jeff Sember—Mattrick developed a game on an Apple II computer. At a time when most computer game developers were focused on making games more difficult and faster-paced, they decided to try something new: a multi-level game based on the theme of evolution. Mattrick approached a local company to get some feedback on the game. He met with the head of marketing and the company president on a Friday, and by the next Monday he had made his first big sale. Within no time, *Evolution* was a Top 10 game, and Mattrick and Sember started to build their company, Distinctive Software, around it.

But Mattrick's mind was still on university. He could have gone to the U.S., but decided that if he studied at Burnaby's Simon Fraser University, he would have more flexibility to build his company. Within two years of Evolution's success, he and Sember had turned Distinctive Software into a company of 25 people. In 1986, when Sember decided to take another path, Mattrick found himself at a crossroads. He bought Sember's equity share in the company and quit school to devote himself to his fledgling business.

Over the years, Mattrick had worked with all of the leading publishers in the PC and video game space and had many opportunities to sell his company or take it public. But, he says, "At the end of the day, I was most interested in picking the company that was willing to be the best cultural fit, that had the most articulate vision for growth, and that was going to win in our space." By 1990 the company was at another crossroads. One path was to focus on building the distribution network for his growing company. The other was to partner with a company that already had a distribution network in place. That was when Mattrick entered the deal with Electronic Arts. "I had worked with them over the years and thought they were the best executive team with the best vision for growth. I said, look, we've got this great studio, we're the largest independent studio in North America. We have 85 people in Vancouver; you've got 200 people in San Francisco. If we stuck our two companies together we could really grow at an accelerated pace."

Electronic Arts agreed. The deal was structured as a pooling transaction because Mattrick was more interested in taking equity in the company than cashing out. It was clear to him that the market was shaking out. There were now 20 companies competing for leadership in the gaming space. When he had started there were 150 companies, "each competing for single-digit market share. But the race was just really starting at that time. And the culture of EA, the people, just the attitude, was in line with the business that we'd built, and we put the two businesses together. It's been a tremendous success since that point in time."

The growth is not just with EA, Mattrick says; the entire industry "has gone through this really accelerated growth curve. If you look at the movie business, what they went through in 80 years, or 100 years, we've gone through in 20. The size of our business is comparable to the film business at box office in North America. Online games will provide alternate sources of revenue and distribution." The future outlook is good for the gaming industry. Kids who played games with Mattrick in the '70s have not turned to other diversions. "People who started gaming when they were 15 are now 35, and playing games is just part of their entertainment time. So we're keeping that audience, and we're adding new people in. Our market is growing."

Playing in the interactive gaming business means winning and losing. "You've got to be comfortable both with successes and failures inside this environment. If you really want to innovate and lead, you have to try some new things. Not everything is going to work, but the flip side is, when new things do work that's usually where you get exponential growth that really accelerates your business and your culture. We really like to win, and we're willing to put in the hard work and emotional energy to win. That sense of pride in competition—that's what our company is about. We're not at our apogee by any stretch."

> **"IF YOU REALLY WANT TO INNOVATE AND LEAD, YOU HAVE TO TRY SOME NEW THINGS. NOT EVERYTHING IS GOING TO WORK, BUT THE FLIP SIDE IS, WHEN NEW THINGS DO WORK THAT'S USUALLY WHERE YOU GET EXPONENTIAL GROWTH THAT REALLY ACCELERATES YOUR BUSINESS AND YOUR CULTURE."**

Mattrick loves to play—even when the stakes are high. "What's really fun about our business is it's kind of a complex 'product and people' chess game, where there are great opportunities and challenges that are not really simple to address." While he believes in working to build consensus and a strong culture, at the end of the day, "You've got to place your bet, so to speak, and make a decision."

How does Mattrick decide where to place his bets? "It's an understanding of the market and common sense, so experience helps. And some of it's just pixie dust." It's hard to describe the creative process, but Mattrick takes a very active role in it, more so than perhaps many executives. "You could describe me as ambidextrous. I am able to work both in sort of a structured executive setting and also to help drive products and product design. And that's really fun for me. One of the rewards in our space is, when you work with a creative team and build something that's never been built before, there's just a huge emotional reward."

In other words, in spite of his huge successes, Mattrick still gets to play. But these days play is on a much grander scale than when he began. "My first product was created working with one person over a summer. Now a normal project team would be about 45 full-time people, working anywhere from probably 16 to 18 months on average. And of those 45 people, you might have a world-class animator, artist, audio person, designer, producer, each of the disciplines. What you can build and the process of how you can build it is dramatically accelerated. Our industry is still learning about all the different things that we can do and experiences that we can bring, and how to immerse people in it and reward people for participating in those experiences. It's really fun."

Mattrick is playing on a different level. The competition is fierce and the stakes are much higher. He says the rewards of winning are extraordinary from a creative standpoint—from personal pride and involvement, and from the calibre of people in the company. He sees himself as bridging the creative and publishing sides of the business, understanding issues like, "How are we going to package and communicate what's fun and unique about this product to consumers?" But he also plays with the executive team, "to talk about our growth vision, our financial performance as a company: what's our corporate strategy; who are we going to acquire, or partner with, to grow our business?"

Despite his astute business acumen and his remarkable achievements (including an honorary doctorate from SFU), Mattrick seems to have retained all of his youthful enthusiasm for the game as well as a remarkable courteousness. He credits his family, his colleagues, and his country for his success. "I really like the idea of anything that's going to support and offer confidence to people to grow businesses inside Canada, because I do believe that that is our future."

THE TRANSFORMER

INNOVATOR_ JEAN MONTY, INFRASTRUCTURE BUILDER INNOVATION_ RE-ENGINEERED BELL CANADA ENTERPRISES INTO A WORLD TELECOM AND CONVERGENCE LEADER

When people think entrepreneurship and innovation, they generally think small startup companies. "Telephone company" is probably the furthest thing from their mind. But reshaping a large company takes just as much initiative and creativity—and a whole lot more strength. In many respects the stakes are much higher. The risks of change can be staggering and the costs of failure immense. At the same time, the risks of doing nothing in a changing environment have been likened to a free ride on the *Titanic*. And in Canada, there is no better example of someone who has taken risks and demonstrated the art of transformation and reinvention than Jean Monty, former CEO of Bell Canada Enterprises. He also played a pivotal role in reinventing Northern Telecom almost a decade ago and more recently helped reshape Bell Canada Enterprises to lead in virtually every segment of the market for voice, data, wireless, and even content.

After leading Bell Canada through the difficult transition to completion, Monty became CEO of Northern Telecom (now Nortel) in 1993. The company was in big trouble, with losses of almost $878 million on revenues of $8.1 billion. Its predicament was simple: no growth. For the previous 20 years, Nortel had focused on digitizing North America's switching technology. It was a great strategy: "It was very simple, and it worked unbelievably well," Monty says. The breakup of AT&T had created the opportunity to grow in the U.S. market by selling equipment to the newly created regional Bell operating companies (RBOCs). But Monty recognised that by 1993 this strategy had run its course. The world was changing and changing fast.

Under Monty's leadership, the business grew; revenues doubled to $12.8 billion, with a profit of $812 million. The company's new businesses overtook its old business in a few short years. Equipment sales to public carriers had basically flattened and accounted for only 26 percent of Nortel's revenues in 1997, compared to 40 percent in 1993. Broadband revenues had tripled and accounted for 21 percent of company revenues in 1997, while wireless revenues had grown more than tenfold, accounting for 22 percent of Nortel's business in 1997. Much of the growth had been fuelled by sales in international markets: revenues from European and other international markets had surpassed revenues from Canada.

The company he left was very different from the company he joined. As Red Wilson, the company chairperson, said, Monty deserved credit for "transforming Nortel into a global growth company. Monty led Nortel through one of the most profound and successful corporate transitions in the industry. Under his leadership Nortel has moved from being primarily a North American manufacturer of central office telephone switches to an organization recognised within the industry as a leader in designing and building communications networks."

Monty rejoined BCE as president and COO in 1997, when the company was still reeling from the impact of long distance competition and the dramatic restructuring that followed. It wasn't a pretty sight: "We looked at the situation and said, We're losing long distance competition, losing long distance market share, pricing is going down, we haven't got enough growth in the other pieces of the business to offset that, and basically realized that Bell can't grow anymore. We have spent an unbelievable amount of energy over the last four years to recreate growth platforms—principally in wireless and broadband, and now in services. Broadening the scope of the company and making Bell a national organization were the key objectives."

MONTY'S RECIPE WAS SIMPLE: "INCREASING THE SCOPE, RE-ENGAGING OUR PEOPLE IN A STRONG FOCUS AND A STRONG DIRECTION OF GROWTH; IT IS SIMILAR TO WHAT WE DID AT NORTEL. FOR THE LAST 10 YEARS OF MY CAREER I HAVE USED A LITTLE TRIANGLE. AT THE TOP OF IT IS CUSTOMER SATISFACTION. ANOTHER POINT IS GROWTH AND THE THIRD IS RESPONSIBILITY. AND IF YOU TRY TO BUILD A STRATEGY THAT DEVELOPS ALL THREE, WHATEVER BUSINESS YOU'RE IN, YOU'LL SUCCEED."

Today, BCE is truly a supercarrier—it leads in virtually every segment of the telecom services market, in voice, in data, and in wireless. It is also stands out among North American telecommunications providers. Last year, industry watcher *Red Herring* remarked, "There's a long distance company on the North American continent that's actually doing quite well. You have to go above the forty-ninth parallel to find it. Montreal-based BCE, Canada's largest telecommunications company, has so far been relatively immune to the economic slowdown that has afflicted the U.S. So far, an effort at diversification has panned out for BCE."[1] And a year later, despite the ups and downs, BCE remains a success story, in large part to the efforts of Monty to diversify beyond the base on long distance services. "Five years ago, 66 percent of BCE's revenues came from local and long distance calls. In 2001 that fell to 41 percent, eclipsed by revenues from emerging products and services." As Monty noted, BCE has always searched for a differentiated and sustainable growth model. "It used to be that you built infrastructures, that was an easy model, but it produced great results—99 percent penetration in Canada, low prices, etc. But with new technology and the shift to technology occurring every nine to 18 months, instead of occurring every decade, that model was flawed, because it basically assumed that the regulator and management of the telecoms monopoly were able to manage these technology shifts. Best-in-class infrastructures, telecoms' infrastructures, and competitive prices are necessary, but are not sufficient anymore to win the minds and hearts of customers."

The quest is on to find the added value that customers will actually want to pay for as networks become a commodity. And while the jury is still out on BCE's convergence strategy, it is has positioned itself as a content provider through strategic acquisitions—Bell Globe Media and BCE Emergis.

In Monty's experience at both Nortel and BCE, one of the keys to managing diverse lines of business is "to put together a strong centre, a very disciplined planning process. It's the glue that keeps the group together and puts initiatives together, cross-group, in order to improve the performance of the overall group." It's a tough balance to strike, the need to meet performance targets on the one hand and to give the "stars" a chance to develop. But Monty understands the need for flexibility, of ensuring that the rules don't stifle innovation. "You have to be mindful that you can't inundate the Emergis management process with 30 BCE executives, the same way that you'd handle a much more complex multi-business environment like Bell Canada."

---

[1] Paul R. La Monica, "Fish or Cut Bait: Dial BCE for Growth," *Red Herring*, July 30, 2001.

While the transformations of Nortel and Bell Canada Enterprises were the result of teamwork and a big talent pool, the successful change of course relied on strong vision, powerful leadership, and exemplary management skill. Monty's recipe was simple: "increasing the scope, re-engaging our people in a strong focus and a strong direction of growth; it is similar to what we did at Nortel. For the last 10 years of my career I have used a little triangle. At the top of it is customer satisfaction. Another point is growth and the third is responsibility. And if you try to build a strategy that develops all three, whatever business you're in, you'll succeed."

While connectivity is BCE's foundation, diversification was a key element of the strategy employed by Monty. He led the company through some bold moves—admittedly with mixed results. But like the entrepreneur in a small startup, risks are an inherent part of the game of transforming traditional businesses. Even those who did not like his convergence strategy had to admire its audacity. Finding that delicate balance between stability and flexibility, between the old and the new, is perhaps what has enabled BCE to adapt and prosper. Of course, not all Monty's bets have paid off, as BCE's decision to cut its losses with Teleglobe show. But even his critics have admired his sense of honour and decision to assume the blame and "fall on his sword."

As for making Canada a place the world wants to do business, many of the same principles apply. The key to productivity is strong infrastructure, whether in terms of communications networks, transportation, or health and social services. But it also requires working smarter. "Canada is on the right page, by putting forward a connectedness agenda. We lead the U.S. in connectivity, and that's the way to get the basic infrastructure in place. However, it's necessary but not sufficient. We also need to use the new tools to improve productivity and efficiency."

Monty is fiercely loyal to his country, to his company, and to his friends, loyal, it has been suggested to a fault. While, in retrospect, many industry watchers claimed he should have seen changes coming (they, of course, knew all along), others have not been so quick to crow. And one did not have to be a fan to be impressed with his dignified exit—quick and clean and without a doubt having the best interests of BCE in mind. Although Monty himself has offered little comment, BCE Chairman Richard Currie insisted that Monty's legacy was a rich one. "I think Jean decided that the headwinds going forward were more than he wanted to withstand. At some point in time, even a person of his inner strength and abilities says it's time to turn over."[2] Clearly, innovators in large companies, like innovators in small companies, have both successes and failures. But the results of his efforts to reinvent two of Canada's largest and most critical companies cannot be underestimated. At the age of 54, he is taking some time to consider his next steps, but regardless of the next challenge he assumes, there is little doubt we have not heard the last of Jean Monty.

[2] "Jean Monty Quits as head of BCE." CBC, April 24, 2002, 17:39:24.

FINDING THE MAGIC SPOT

INNOVATOR_ ANTOINE PAQUIN, SERIAL ENTREPRENEUR INNOVATION_ BUILT NUMEROUS STARTUPS, SELLING THEM TO COMPANIES SUCH AS CISCO AND CONNEXANT FOR CLOSE TO A HALF BILLION DOLLARS

Antoine Paquin found the "magic spot" before anyone else and was successful beyond his wildest dreams. It was in 1996, and he was in the right place at the right time. He got into fibre optics, the high-speed transmission medium, before anyone else. The opportunity he identified was very specific. He set up Skystone System to supply semiconductor solutions to equipment manufacturers in the burgeoning communications sector, specifically for transport over fibre optic networks. He says, "We achieved a very quick sales ramp. Within two years our largest customer, Cisco systems, stepped up to the plate and determined that it was strategic to have us in-house, and they acquired the company outright for $120 million (U.S.)."

Paquin was first catapulted into the big leagues by an innocent decision. He took a vacation. "My first startup in 1994 was a consulting outfit that designed telecommunications equipment. The company was profitable, growing through its own sales. In the summer of 1996, I made the mistake of taking a vacation and on that vacation I paid a visit to a colleague in the industry, Greg Aasen, the founder of PMC Sierra Wireless. I visited his factory, where they were testing semiconductor devices, and he explained to me his business model. I remember going back to where we were staying that evening, and thinking to myself, "I've been doing it wrong all along. I've been playing it safe; I've been growing conservatively. Taking risks, but very minimal risks. And really not pushing my full potential." That visit also led Paquin to see the potential at the intersection between the semiconductor and optical transport businesses—the "magic spot." "So I started writing a business plan—the rest is history."

And it is quite a history. Becoming a multimillionaire within 10 years of graduating would have been enough to put Paquin in the Canadian Entrepreneurs' Hall of Fame, but he was just getting warmed up. He has repeated this approach of identifying magic spots with other companies. His knack for spotting new opportunities before anyone else is uncanny. He owns up to an ability "to recognise ideas that are at the forefront of the next phase of an industry—the intersection point between different disciplines. For example, Skystone was the intersection point between fibre optic transport and data communications." After the acquisition by Cisco, he became their director of business development, in charge of the fibre optics space. At the same time he acted as initial investor in various companies in Ottawa, and ended up running one of them, a company called Philstar, starting in November 1998. Philstar was the intersection point of data communications and wireless. A couple of years later he sold it to Connexant Systems for $250 million (U.S.). In January 2001, he joined Bit Flash as CEO. Paquin sees Bit Flash as the intersection point of graphically rich, intuitive applications and content applied to the mobile space, a market that doesn't really exist yet.

To find the magic spot in an emerging industry, Paquin talks about imagining for people what they don't know they're going to need. Illustrating his point, he takes us back to 1984, when an Apple IIC with wordprocessing and e-mail was considered leading-edge. "Now you've become accustomed to intuitive, graphically oriented applications and content on your desktop. If I tried to remove that, I'd have a revolt on my hands. But if I asked you if you wanted it in 1984, you would have said, 'Don't be ridiculous,' because you didn't know you needed it. That's exactly where wireless is today; that's why it's a good time to get in."

Paquin's success, of course, does not rest only on his ability to see into the crystal ball. He has also been very successful in getting others to buy into his vision. "I've always used external capital. My first company was bootstrapped (with no outside capital). So I've experienced both ends of the spectrum; that is, growing through your own merits or growing through capital infusion. And I can tell you that the discipline that's involved in bringing outside capital is a good discipline to undertake for any startup. Of course you can go to an extreme, but the discipline that knowledgeable, value-added, sophisticated asset investors bring to your company is the desirable one."

ALTHOUGH PAQUIN IS OVERLY MODEST, HE DOES ADMIT TO SOME UNIQUE STRENGTHS THAT HAVE HELPED HIM FIND "THE MAGIC SPOTS." "I THINK BOT-TOM-LINE. I HAVE A GOOD NOSE FOR OPPORTUNITY. A VERY OBJECTIVE MIND-SET. A CLEAR SENSE OF VISION AND AN ABILITY TO RECOGNISE TALENT IN OTHER PEOPLE." HE ALSO ACKNOWLEDGES THAT PERSONAL DRIVE IS A VERY IMPORTANT CHARACTERISTIC.

He is clear about what made his companies attractive to venture capitalists. "Basically, we allowed them to make money. It was quite clear that these were big opportunities, if they were exploited properly, and that's the only thing needed to satisfy a venture capitalist—that and whether or not you have the right team to execute them. And of course we had a first-mover advantage. Every company I've been involved in was a first mover in its market space."

AND PAQUIN'S REMARKABLE SUCCESS HAS GIVEN HIM A KIND OF FREEDOM THAT HAS LITTLE TO DO WITH MATERIAL SUCCESS BUT IS A STATE OF MIND. "THE DAY YOU REALISE THAT YOU HAVE ABSOLUTELY NOTHING TO LOSE, THAT YOU EXIST IN A SOCIETY THAT REWARDS RISK-TAKING AND FORGIVES FAILURE, IS THE DAY YOU MAKE THE TRANSITION. WE'RE BLESSED TO LIVE IN A SOCIETY THAT VERY MUCH REWARDS SUCCESS AND IS VERY FORGIVING OF FAILURE, AND IF ANYTHING HAS LEARNED TO RECYCLE FAILURE INTO SUBSEQUENT SUCCESS."

Although Paquin is overly modest, he does admit to some unique strengths that have helped him find "the magic spots." "I think bottom-line. I have a good nose for opportunity. A very objective mindset. A clear sense of vision and an ability to recognise talent in other people." He also acknowledges that personal drive is a very important characteristic. "It takes a big ego to do that kind of work, but you want to have somebody who has a big ego that is under control. That's a very important aspect, and that tends to lead to other results, such as teamwork and co-operation. Big egos can make companies. They can also destroy them."

"What makes companies successful is the ability to set a direction, stick to it, but have a very direct feedback loop into the organisation in terms of what's going on in the marketplace, and adjust the plans until you hit that magic spot where all of a sudden you've found a solution to a problem that a customer couldn't even tell you they had. They knew they had a problem; they just didn't know quite what the problem was or what the solution was."

Paquin was educated as an electrical engineer and, like many in the industry, got his start as a communications chip designer at Bell Northern Research in 1989. What set him aside from his classmates? In a word, impatience. "I didn't see myself growing in a big company and I always had a fiercely independent mindset, so at the end of the day, I think the entrepreneurial path was inevitable," he says. As with many other entrepreneurs, entrepreneurship ran in Paquin's family. And Paquin's remarkable success has given him a kind of freedom that has little to do with material success but is a state of mind. "The day you realise that you have absolutely nothing to lose, that you exist in a society that rewards risk-taking and forgives failure, is the day you make the transition. We're blessed to live in a society that very much rewards success and is very forgiving of failure, and if anything has learned to recycle failure into subsequent success."

This highly adaptive business model works especially well in the Canadian context. "I think Canada is an excellent place to build a company, which is why I'm still doing it here. Canada has a lot of talent. A lot of engineering talent, a lot of creative talent. And when we develop technology, it is all for exports—Japan, Europe, Korea, the rest of Asia, the United States, South America. Everything we design is meant for an export market, as we don't have a big local market. And I think it's both a weakness and a strength. It's a weakness because your early markets are deployed offshore, not locally, and typically for an early market you like to be very close, physically, to the customer. We can't do that, so we have to be very close in responsiveness, but not in physical proximity. And it's a strength because ultimately we tend to adopt a global perspective of our product in our solution. Canada is also very stable— it has a stable legal system. Stable currency. Stable politics." In a rapidly changing world of changing technology and changing markets, Canada's stability and dependability are assets.

JESSE RASCHE "I always surrounded myself with great people. If a group of great people put their minds to something, they can do just about anything they want."

CASH, EQUITY, AND COMMUNITY... CANADIAN STYLE

INNOVATOR_ JESSE RASCHE, SERIAL ENTREPRENEUR, FINANCIER INNOVATION_ BUILT AND SOLD 51 PERCENT OF INQUENT, A WEB-HOSTING WHOLESALER, FOR $115 MILLION (U.S.) TO SBC COMMUNICATIONS IN TEXAS

Jesse Rasche made headlines in 2001 when, at the ripe old age of 25, he sold 51 percent of InQuent Technologies Inc. to Texas-based SBC Communications Inc. for $115 million (U.S.). The story of how he created a market for Web hosting and exploited it is quite unique. But even more noteworthy is Rasche's uniquely Canadian recipe for success. Like many entrepreneurs, he attributes his success to three personal characteristics: persistence, creativity, and salesmanship. But his strategy for building a successful company rests on recognising an important balance, the balance between cash and equity on the one hand and community on the other. Investing in and supporting human capital and developing community in his companies and in Canada is as important to Rasche as hard cold cash. And investing in community pays off in many ways.

Like many entrepreneurs, it runs in the family—his father is an architect and his mother a real estate agent. Rasche got his start in his teens by setting up a real estate A-frames for a couple of home builders. By the end of the summer he had 40 to 50 clients and eight employees. His next venture, at the age of 19, was working for a natural gas wholesale distributor. Again, within months he had grown the business and had 200 people working for him.

Rasch decided it was to time to get "an education." "When I went to McGill University, I told my mother, I'm going to do my best to focus, but if an opportunity comes up I'm going to seize it and make the best of it. And that's pretty much what happened. I kind of broke a promise to her that I wouldn't drop out, but I did. We started exploring the Web hosting business in 1996, and I dropped out at the end of 1997, in my third year.

"I recognised at the time that Web hosting was in its infancy, and was growing very quickly because small businesses all around the world required Web site hosting in order to get their sites on the Internet. The other services came along with that." So Jesse established InQuent. "What we did, that made us different, was create a very good user interface, that was accessible on the Internet, that allowed small businesses with no Web savvy or technology savvy to very easily manage their sites, their company e-mails, and everything that has to do with their online Web presence. We just put up a page on the Internet, and before we knew it we were getting orders from all around the world. And when those orders started trickling in without any real marketing or advertising, we realised we were on to something pretty big.

"I always surrounded myself with great people. If a group of great people put their minds to something, they can do just about anything they want." So attracting and retaining human capital is essential. "I had the benefit of having access to an immense amount of IT talent in the Montreal area, students from McGill University and Concordia, Computer Science, who could very cost-effectively help us create the technology infrastructure for our business—much more cost effectively, I should say, than our American counterparts. One of the greatest benefits of doing business in Canada is certainly the affordability of Canadian talent, whether it be management talent or IT talent. We are billing our customers in U.S. dollars and paying our employees in Canadian dollars."

In next to no time, 95 percent of Rasche's business was international. He notes, "In the end, we were operating in over 140 countries, when we were in the retail business of Web hosting." However, Rasche assessed the emerging competition, and the company shifted emphasis to deal exclusively with the wholesale end (in effect providing the back end for branded Web hosting services offered by telecommunications providers—companies like Bell Canada and Telecom New Zealand). "We were the first company on the planet to do this. We essentially invented an entire industry, which was wholesale Web hosting."

> **LIKE MANY ENTREPRENEURS, HE ATTRIBUTES HIS SUCCESS TO THREE PERSONAL CHARACTERISTICS: PERSISTENCE, CREATIVITY, AND SALESMANSHIP. BUT HIS STRATEGY FOR BUILDING A SUCCESSFUL COMPANY RESTS ON RECOGNISING AN IMPORTANT BALANCE, THE BALANCE BETWEEN CASH AND EQUITY ON THE ONE HAND AND COMMUNITY ON THE OTHER. INVESTING IN AND SUPPORTING HUMAN CAPITAL AND DEVELOPING COMMUNITY IN HIS COMPANIES AND IN CANADA IS AS IMPORTANT TO RASCHE AS HARD COLD CASH.**

He goes on, "Although telephone companies generally move at the pace of dinosaurs, when they started to sniff around our industry we knew that it would only be a matter of time. They are the largest service providers to our target market—small and medium-sized businesses. There's no reason why, if they went into the market, they wouldn't crush us pretty quickly if we didn't have a very strong brand. So we thought about it strategically; if we were going to compete with the telephone companies we would have to spend tens of millions of dollars, branding the business and marketing it, to create a formidable presence. And even then, we would never be able to compete with this inherent inertia that they have with their own customer base with their easy bundling and marketing. So we said, we could either try to compete with them and spend everything we have on building a brand, or we can partner with them, access all of their customers, and still grow our business. And we said, 'You know what, let's partner with them.' So we said to them, 'Don't buy us; let us sell you our products. We'll create your entire back end infrastructure and you can just private label, or resell, what we have.' We were the first company to do that."

> "I RISKED A LOT. BUT I FIGURED, YOU KNOW WHAT? I'M YOUNG ENOUGH, IF I HAVE TO DECLARE PERSONAL BANKRUPTCY IT WON'T BE A BIG DEAL. YOU NEED TO BE ABLE TO TAKE RISKS AND FOLLOW THROUGH."

Of course, it's not all been easy, or a license to print money. "There are a lot of roller coasters when you're an entrepreneur. You have to be very persistent in trying to reach your goal. When we started this first business, InQuent, I completely leveraged myself. I had credit agencies after me; every credit card was maxed out. I took my student loans and put them into this too. I risked a lot. But I figured, you know what? I'm young enough, if I have to declare personal bankruptcy it won't be a big deal. You need to be able to take risks and follow through."

Rasche firmly believes that community is as important as cash and equity. Money alone is not enough to build and retain a team. His priority has been "surrounding myself with incredible talent and creating an environment where everyone feels like they're family. Quite frankly, when we were starting the business and running it for the first three years, before the SBC transaction, we worked 12- and 16-hour days, every day of the week. Money alone would not provide an incentive. You had to sell other people on the opportunity to be part of a family, literally, because they would be spending so much time together. And we created a tremendous social network. I think people have got married, several of them, who met each other at our business, and it was a lot of fun. Creating that social dynamic just made people so passionate about working for the business that they didn't consider it to be work. So creating an environment where that kind of passion was present was of paramount importance to our success."

Rasche adds that part of building community, in the broader sense, is investing in health, education, and the environment, applying what he has learned to the non-profit sector. Today Rasche splits his time between his commercial interests in Aprilis; his private equity business; and VerticalScope Inc., a leader in delivering specialized content resources online for the SME market. "The other half of my life is spent on venture philanthropy. It's an approach to philanthropy that's modelled on venture capital. It takes the same principles of requiring charities to stick to their plans, develop really unique ways of approaching a social problem in their market, and being really innovative in terms of how they address those issues." Rache, only 26, has established a foundation to fund philanthropic projects on this model.

And he continues to wax poetic about his home and native land. "My advice to Canadian CEOs on stemming the flow of talent to the U.S. is to be persistent and not to hesitate in waving the Canadian flag. There is a lot of pride among Canadians about helping Canadian companies become more competitive in a global economy. Remind them that Canadian cities are safe, that this is a great place to raise children, that there's a highly rated educational system. All these things are part of the package."[1]

It's little wonder his mother has forgiven him for dropping out of university.

[1] Heidrick & Struggles Leadership Opus, 2001.

ANDREA REISMAN JOHNSON "There will be random moments in time, moments you're not expecting, when you catch someone at a cocktail party or see someone in a restaurant. And you need to have, on the tip of your tongue, a succinct way to explain what you're doing and get somebody interested. It's the business equivalent of a movie trailer." If the trailer is good, people want to see the movie."

THE WINNING
PITCH

INNOVATOR_ ANDREA REISMAN JOHNSON, SERIAL ENTREPRENEUR INNOVATION_ FOUNDED PETOPIA.COM, SETTING OFF A TWO-YEAR EXPLOSION OF PET PRODUCTS SOLD OVER THE INTERNET

One of the things budding entrepreneurs learn about in business school is the "elevator pitch." The premise is that an entrepreneur finds herself in an elevator with a venture capitalist and persuasively describes her business idea before the VC gets off the elevator. While Andrea Reisman Johnson acknowledges, "I've yet to be in a scenario when you have 30 seconds with someone to explain your life's story," she has had many, many opportunities to hone her pitching skills. As the founder and CEO of Petopia.com, Johnson created opportunities to pitch her business ideas. The end result? After at least 90 meetings, with 40 different venture capitalists, at nine firms, she raised $75 million (U.S.) for Petopia, an online pet store aimed at tapping into the $23 billion pet industry. What are the secrets of Johnson's success? Be prepared. Be persistent. Work hard. And above all, continuously improve your pitch.

Johnson got her start at Canadian beverage maker Cott. Under her management, the startup Alternative Beverages division grew from revenues of $3 million to nearly $100 million. After five years at Cott, Johnson went to Harvard, where she earned her MBA (and ran the Boston marathon a couple of times).

Although she is a Canadian through and through, Johnson wanted to spread her wings. "I wanted a chance, as I think most entrepreneurs do, to try something on my own. I'm lucky to have a terrific family, but also a fairly well-known family. I really wanted a chance to go out on my own and pitch investors who had never heard of me and who had no reason to give me the time of day."

> **WHEN THE ENTREPRENEUR UNDERSTANDS HOW THE DECISIONS ARE MADE AT A PARTICULAR FIRM, THE GAME BEGINS IN EARNEST. THE KEY IS TO MAKE SURE THAT EVERYONE WHO HAS A STAKE IN THE DECISION HAS AN OPPORTUNITY TO HEAR THE PITCH, AND THAT EVERYONE WHO HEARS IT RETAINS THE MAIN MESSAGES.**

She soon got her opportunity. After a chance meeting at La Guardia airport in New York, Johnson partnered with high school friend Lorne Abony to develop an online pet company. A pet lover herself, Johnson knew that there were opportunities to shift some of the $23 billion pet industry from traditional to online channels. When she and Abony started to look for financing for Petopia.com, she was able to draw upon her business school contacts to set up initial meetings. But she wasn't content to rely upon her contacts alone, so she also did a lot of cold calling—and learned how to hone her skills. As CEO of Petopia, Johnson secured several rounds of VC financing. Petopia didn't survive the dot-com crash as a standalone company, and in early 2001 to the company was sold to bricks-and-mortar Petco for an undisclosed amount. Petco had invested in, and collaborated with, Petopia, and it now operates the site as its own online retail division.

During the process of building the company Johnson learned a great deal. While she has yet to pitch a VC in an elevator, she firmly believes that the principle of the elevator pitch is sound. "There will be random moments in time, moments you're not expecting, when you catch someone at a cocktail party or see someone in a restaurant. And you need to have, on the tip of your tongue, a succinct way to explain what you're doing and to get somebody interested. It's the business equivalent of a movie trailer." If the trailer is good, people want to see the movie. Likewise, if the initial contact is interesting, it opens doors for an entrepreneur to present a detailed business idea to a VC firm. "The point is not that they're going to write you a cheque. The point is to have them agree to give you half an hour of their time to listen to a deeper story."

Once an entrepreneur has secured that half-hour of VC time, it's crucial to make a good impression. It's not nearly as easy as it might seem, as Johnson explains. "You have to understand the process in each firm. They all have different processes for approving investments. Some will say, 'All you need is one partner to be in favour,' and that person can champion the deal and make it happen. Other venture firms will say, 'It's majority rules.' Other venture firms will say, 'It's unanimous, everybody has to have seen the pitch.'"

When the entrepreneur understands how the decisions are made at a particular firm, the game begins in earnest. The key is to make sure that everyone who has a stake in the decision has an opportunity to hear the pitch, and that everyone who hears it retains the main messages. "In your mind you have a scorecard. For example, there are six people at the table and there are seven points to get through. You must keep track of each individual and the points they've absorbed. In the end you need to know that you checked all seven boxes for all six people."

To make this "mental grid" manageable, "it's important to boil down your points. You need to figure out what your key points are. And then you keep track, so you know where each person is at all times." And it's not just key points that Johnson tracks; she's also acutely aware of the energy level in the room—who is paying attention and who isn't. "People will get up and leave at various intervals. The senior people are pressing on the junior people. And somebody else's cellphone is always ringing. There's all kinds of stuff going on, but you stay confident and you stay focused. And make sure you do other things. Make sure to engage them; you can't just talk for 10 minutes, because people will tune out. It's important to have various questions throughout the pitch where you bring them in. It's a hard balance, because you don't want to lose the flow, but you have to bring them in."

AS JOHNSON DEMONSTRATES, IT'S NOT EASY TO DEVELOP A GOOD PITCH. IT TAKES LOTS AND LOTS OF PRACTICE AND PREPARATION BEFORE ENTERING THE ROOM. AND ONCE IN FRONT OF THE AUDIENCE, THE "PITCHER" NEEDS TO CONTROL THE ENTIRE PROCESS, AND ENSURE THAT ALL THE KEY DECISION MAKERS GRASP THE ESSENTIAL PARTS OF THE STORY.

It's evident that Johnson is a major-leaguer when it comes to pitching her ideas. She's constantly reading her audience, checking her timing, and staying "on message." It's easy to be distracted by questions, but Johnson has developed a routine to make sure she delivers her message according to her plan, in the time available. "Ask up front, 'How much time do you have?' Then tell them up front, 'I am going to go through the whole pitch. It's going to take me this much time. I will answer all the questions afterwards. Feel free to stop me in the middle, so I can make a list of the questions, but I'm going to answer all the questions afterwards.' To the extent that someone forces you off track with a question, you obviously can't be rude to them. Answer the question and then bring them back to the pitch. Make sure the pages on the pitch are numbered. As silly as it sounds, you can say, 'Mr. So and So, you asked a question. Here's the answer to your question, now let me take you back to page 3 of the deck.' You can't avoid any of their questions, but you need to get through your story. Their questions will actually make more sense once you have given them some context."

The job is not complete until the loose ends are tied up. "If you're in a place where the decision needs to be unanimous, and there are 14 partners, then you have to note how many people you've met. You need to ask, 'How do I get to meet the other partners? What are the next steps?' This is hard, because the people at the table want to tell you that they're the only ones who are important. You need to build relationships with them while finding out who else is involved in the process."

As Johnson demonstrates, it's not easy to develop a good pitch. It takes lots and lots of practice and preparation before entering the room. And once in front of the audience, the "pitcher" needs to control the entire process, and ensure that all the key decision makers grasp the essential parts of the story.

And although there is little question that a good pitch is as important today as it was when she started, the Canadian venture capital market has clearly evolved since her trial by fire with Petopia. "I feel really strongly about Canada. And so at some level I've got mixed feelings about leaving." But, "at the time we were looking to build [Petopia], Canada wasn't ready to fund us to the tune that we needed to be funded. As a result, we really wouldn't have been likely to secure the Petco relationship, and as a result of that, we would have had a whole series of other challenges," says Johnson. "In the last few years the environment for young businesses has changed in many ways on both sides of the border. Some of the changes are positive for Canada. Today, Canada's venture market is proportionately sized relative to its U.S. counterpart. (The Canadian population is roughly 10 percent that of the U.S. and the Canadian venture pool is roughly 10 percent that of the U.S.) Additionally, relative to their American counterparts, Canadian venture capitalists are more comfortable funding seed stage companies. If I were a Canadian-based entrepreneur building a business now, I'd start the business at home, in Canada, and wouldn't look for U.S. funding unless it was needed in later-stage, larger rounds."

MADE IN CANADA.
SOLD IN THE USA.

INNOVATOR_ EMAD RIZKALLA, SERIAL ENTREPRENEUR INNOVATION_ FOUNDER OF WEB APPLICATION AND E-LEARNING SOLUTIONS PROVIDER ZEDDCOMM, DRIVING THE COMPANY TO 1,300 PERCENT GROWTH IN FIVE YEARS

Conventional wisdom suggests that when you start a new business, you start small. But Newfoundland-based entrepreneur Emad Rizkalla did just the opposite. His company, ZeddComm started big, because it couldn't afford to start small. Right from the start Rizkalla knew that a solid base in the U.S. was essential. Today ZeddComm is servicing organizations as diverse as Cisco Systems, the NBA's Golden State Warriors, and Health Canada. The company designs, develops, and deploys custom Web-based business applications and e-learning solutions from its four office locations (Orange County, San Jose, Ottawa, and St. John's).

Rizkalla explains the standard approach to building a new business. "The traditional model is that you start a business focusing on a local market, and you succeed by convincing local businesses to buy your products. And then when you feel really adventurous you'll go to Halifax and try to convince someone there to buy it, and then when you're feeling really, really brave you'll expand to Moncton. That is conventional thinking on the way a business is supposed to grow. But really, that was never our approach. From day one, our approach was focused on building credibility because we were a couple of kids in the middle of nowhere, with no money, no experience, and no contacts. It wasn't 'normal' in 1992 to be a 23-year-old technology entrepreneur, so we couldn't afford to grow using the traditional model. We didn't even have enough credibility to grow our base in the local market. We also didn't think we had enough time or money to do it that way, so we headed straight for the U.S."

BUT TECHNOLOGICAL EXCELLENCE IS NOT ENOUGH. RIZKALLA CAUTIONS ENTREPRENEURS THAT SUPERB TECHNOLOGY ALONE DOES NOT GUARANTEE SUCCESS. "YOU HAVE TO BE OBSESSED WITH UNDERSTANDING YOUR CUSTOMERS—WHAT THEY WANT, AND WHY THEY WANT IT. THIS UNDERSTANDING SHOULD FORM THE BASIS OF YOUR VALUE PROPOSITION." AND A VALUE PROPOSITION IS MORE THAN GOOD TECHNOLOGY, IT'S ALSO ABOUT UNDERSTANDING THE MARKET.

Starting big allowed ZeddComm to grow by 1,300 percent in five years, and in the past three years, the company has appeared twice on the prestigious Profit 100 list. As Rizkalla says, "We were from a place where there wasn't a great deal of opportunity. However, we knew we had great talent, and we had great thinking, we had very good technology, but we had no market. So we went to a place that has an infinite market. It is a great example of supply and demand." Indeed, Rizkalla is so passionate about starting in big markets that in his role as chair of the Newfoundland and Labrador Association of Technology Industries, he constantly challenged small companies to expand their horizons and look beyond their local markets.

More than 60 percent of ZeddComm's revenues come from outside Canada. Rizkalla says, "We now make significantly more profit than we've ever made focusing on the local market." But he also notes that about 80 percent of custom development (or the "value add") is completed in Newfoundland, by the "great talent" mentioned above. As an adopted Newfoundlander (he's lived there since he was 12), Rizkalla recognises that it's Newfoundlanders' strong ties to home that give ZeddComm part of its competitive advantage, far away from the technology clusters of Silicon Valley, Ottawa, and Waterloo.

"People from Newfoundland have a very strong attachment to place. People who leave—and obviously, if you know of any Newfoundlanders, you know exactly what I'm talking about—typically leave here very reluctantly and are usually anxious to return. So that allows us to bring talented people home. Some of these individuals have exceptional credentials and excellent experience. Most are people who would normally never end up in a smaller company, in a place like this, but for the fact that they're from here and they really want to come home. In terms of access to talent, places like Newfoundland would surprise you. If you look at our staff, and some of the places that they've worked, some of the places that they graduated from … it is remarkable. There are very few companies of our size in Los Angeles or Ottawa that could boast that they employ graduates from Oxford, Berkeley, and Harvard. However, what impresses our clients at the end of the day is the obsessiveness of our employees with delivering customer service. Our staff love what they do, and love the fact they can work on challenging projects, for the world's most demanding clients, from 'home'—the place that they love."

In fact, Rizkalla wishes there were more incentives to help companies recruit and repatriate top talent. "In this whole sector, whoever has the best people wins. And if you look at job creation in the sector, you always leverage the more experienced individuals to create three or four positions for people with less experience; that's what drives the sector. The senior people add the bulk of the vision, talent, and leadership in the various technology areas. So I think it would be strategic if the various governments would focus on programs to bring top Canadian technology professionals back home. We need a program (perhaps a short-term tax credit) that would make it more enticing for individuals with specialized experience to relocate back to Canada, because that's really what's going to give us the edge in international markets."

Repatriating talent and exporting products and services have proven to be very effective strategies. It's the best of both worlds—access to the U.S. market and associated U.S. dollar revenues, created with Canadian expertise paid in Canadian dollars. Not only are there cost advantages in developing custom solutions in Canada, Rizkalla thinks that Canadians have more loyalty to their employers. "In our knowledge-intensive industry, a company's offering is directly a function of its people, so loyalty and lower turnover are huge competitive advantages. This is an example of a solid, tangible competitive manifestation of a cultural difference between the United States and Canada."

ZeddComm uses its made-in-Canada technology expertise to serve U.S. markets. Rizkalla admits that there are some stereotypes about Canadians in the U.S., but he argues that they're positive. "The fact that you're generally likeable, and you're from a laid-back place that's tolerant and friendly, is an advantage. Canadians need to leverage some of the perceived stereotypes about being very friendly and polite. People want to do business with people they like, so why wouldn't you leverage that? The trick is to build an awareness of our technical capabilities on top of these underlying perceptions. We need to demonstrate continually that Canadians also have a focus on developing technology that is superb and that we have provided innovations that have been every bit as impressive as many of those originating from the U.S."

But technological excellence is not enough. Rizkalla cautions entrepreneurs that superb technology alone does not guarantee success. "You have to be obsessed with understanding your customers—what they want, and why they want it. This understanding should form the basis of your value proposition." And a value proposition is more than good technology, it's also about understanding the market.

"One of the quintessential mistakes people make when they're starting a company is that they allow their passion for what they're doing, or what they're developing, especially in the technology sector, to overwhelm their attention to potential customers and how the marketplace is going to perceive their offering. I've seen too much focus—way too much—on technology above all else, and this is not a luxury that small companies can afford. From day one, they have to be market-oriented. Technology for ZeddComm is a means to an end; we must ultimately solve a real-world problem for our customers if we want to survive and grow. That is what drives us, not a single-minded passion for technology."

Rizkalla shares an anecdote he heard from a venture capitalist. This VC has many entrepreneurs approaching him for funding. "Many of the small companies were frustrated because the VC was not jumping up and down to bestow his hard-earned money upon them. And one of the entrepreneurs told him, 'You know, if I only had a million dollars, I know I could sell this product. I know I could make it successful. I know customers would want to buy it.' And what the VC told me, and what he told this chap, was, 'I'm your first customer. If you can't sell me, you're not going to sell it to anybody else.' So even financing can be directly linked into the framework of market orientation."

Entrepreneurs need to have passion to succeed, but Rizkalla cautions that this passion can be misdirected. "I've given advice to a lot of technology startups, and they're really excited about what they're doing, they have all the passion that they need to succeed, but the passion is misplaced. It's not a passion in what you're developing, I think, that makes you successful. It's a passion in knowing how your offering is going to help people, or help companies. If you don't have that passion for helping your customers, you're not going to make it. But if you do have it, I think everything else can fall into place."

Rizkalla focused his own energy and passion into starting big. For ZeddComm, the big market was not at home, but in the U.S. But the talent was at home, and by adopting a made-in-Canada, sold-in-the-U.S. approach, the company has prospered.

SKATE TO WHERE THE PUCK WILL BE

INNOVATOR_ TED ROGERS, INFRASTRUCTURE BUILDER, SERIAL ENTREPRENEUR INNOVATION_ FATHER OF ONE OF CANADA'S LARGEST MEDIA AND COMMUNICATION EMPIRES

It's been more than 40 years since Ted Rogers started his first company. As a law student he parlayed his love of music into a successful orchestra supply company that also sold Polaroids of couples on the dance floor. His first effort, combining music and image, oddly prefigured his creation of one of the largest media and communications companies in Canada. With fingers in virtually every segment of the convergence pie, including radio and television broadcasting, high-speed Internet, wireless, and "content," Rogers Communications Inc. (RCI) has become a dominant force in Canada.

Rogers has been remarkably successful at anticipating changes. "The key is the trends, not the results of any specific day. And if we've had any success at Rogers, it's because we've been able to get ourselves not just in line with trends, but ahead of them." Looking at his remarkable career, some would say Rogers has shown an uncanny ability to skate to where the puck will be.

Edward S. Rogers, Jr., pays homage to his remarkable father, Edward S. Rogers, Sr., at every opportunity. The electrical engineer was the inventor of the first alternating current (AC) radio tube, enabling radios to run on ordinary household current. He went on to establish Canada's first all-electric radio station (CFRB: "Canada's First Rogers Batteryless") and later received the first private television license in Canada. He died in 1939 at the age of 38, but his legacy is all around us, not only in the huge donations made in his name by his son but in RCI's very success.

> "THE KEY IS THE TRENDS, NOT THE RESULTS OF ANY SPECIFIC DAY. AND IF WE'VE HAD ANY SUCCESS AT ROGERS, IT'S BECAUSE WE'VE BEEN ABLE TO GET OURSELVES NOT JUST IN LINE WITH TRENDS, BUT AHEAD OF THEM." LOOKING AT HIS REMARKABLE CAREER, SOME WOULD SAY ROGERS HAS SHOWN AN UNCANNY ABILITY TO SKATE TO WHERE THE PUCK WILL BE.

Rogers, Sr., learned a lot from inventors in approaching innovation. Ted Rogers notes, "My father visited all sorts of laboratories. He talked to all sorts of people, he read all the time, and he experimented in developing the alternating current tube. I have done the same in order to understand where the market is going."

Ted Rogers moved from his orchestra supply startup to founding CFTO-TV and purchasing CHFI-FM radio. His entry into broadcast television and radio was inspired partly by his passion for music and partly by his father's work—and driven by his instinct for trends. "I was extraordinarily interested in music. I just honestly believed that high fidelity and stereo were going to change things, and that FM was the type of radio that would succeed in the future." His early gamble on FM paid off and began a career marked by an uncanny ability to be where the trends are going.

"Cable may seem obvious now, but in 1965 it wasn't, " he says. His idea was radical for the time. "The idea that the future was not large-scale broadcast TV stations, with homogenized programming, the news at a certain hour, sports at a certain hour, and *I Love Lucy*, and then a movie. … I believed that you would have whole channels devoted to movies or sports or news. There weren't enough channels, enough spectrum, to handle that. Cable was the solution."

He uses the same logic to guide his assessment of new opportunities at RCI. He has good instincts. "Last year, for example, we did not think VOD [video on demand] would proceed within the next 12 to 18 months to any great degree, because the movie studios were not energetically behind it. This year there's a sea change. The movie companies are now moving toward making their programming available for this." What he describes is a combination of formal collection and analysis of information coupled with an instinct for seeing what will come next. Like Wayne Gretzky, skating to where the puck will be.

There are no mysteries in the trends that Rogers sees shaping the world—globalization, increased competition, and technological change. Video on demand is one example of the way technology has changed, enabled by increased intelligence and two-way capability in the network: "The public wants choice. They want value." But Rogers saw this coming early; his company began more than a decade ago investing heavily to make 89 percent of its network two-way. This provided a solid infrastructure on which to develop applications that were largely unimagined in the years before the explosive growth of the Internet.

"Television has high penetration, is familiar technology, and is easy to use. Our new interactive television services support new applications, such as gaming and video on demand. High-speed Internet access will remain an important market, despite initial growing pains. The company is continuing to invest heavily to stay ahead of the demand. What else might we expect in the future? It comes down to good old-fashioned common sense. The basic test is always the same: 'Do you have services customers want? Can you keep them satisfied? How much will they pay?'"

In terms of managing his business, Rogers stresses the importance of bringing together a variety of perspectives to assess trends and develop strategic frameworks. "As the rate of change accelerates, the importance of taking time out and doing strategic planning, of trying to figure where we want to go, and where we're being taken, is even more important. You must be willing to do battle within the company. Everybody should be free to speak. You fight like heck for what you believe in. And you should be willing to change. This year, I've started out on one side of several issues, and then ended up on the other side, because I realized that that made sense."

> **"AS THE RATE OF CHANGE ACCELERATES, THE IMPORTANCE OF TAKING TIME OUT AND DOING STRATEGIC PLANNING, OF TRYING TO FIGURE WHERE WE WANT TO GO, AND WHERE WE'RE BEING TAKEN, IS EVEN MORE IMPORTANT."**

Ultimately it is the numbers that matter. "I think it's important for a strategy that it be economically viable and technically feasible, as best as you can determine. Once we get everyone to agree on the strategy, they are freer to implement that strategy without needing to get all of us together, week after week and month after month, debating whether to do this or do that." The key to this is a diverse pool of talented people who will challenge the boss. "You want to get people around you that are younger—you don't want everybody the same age—and smarter, because you don't want a bunch of dummies around. You want people who will question you and challenge you."

Rogers thinks that his law school education pushed him to test the evidence, to focus on the facts. And in a game where mammoth egos often prevail, Rogers is a firm believer in the value of constructive criticism, even when he is the target. "Pure entrepreneurs are unpredictable, and often think they're smarter than everybody else. They may have had success, but that attitude often leads to their failure. They hit three home runs in a row, they get up and they assume they're going to hit a fourth, and that's when they strike out and lose everything. They've hired some yes-men around them, and they haven't got people who say, 'Ted, you know, you're a bloody fool.' I don't mind if they're saying, 'You're a bloody fool.' I say, 'Well, now, what's the evidence of that? You give me your numbers, I'll give you mine.' And we have these debates all the time. You have to work very hard as an entrepreneur. You have to believe in yourself, but also look in the mirror occasionally and realize who you are. Not perfect."

Many of the lessons that apply to entrepreneurs also apply to Canada itself. "We have a great education system and a great pool of talent. One advantage of being in a smaller country is that the people know each other a lot better, and that's beneficial. People are gentler, kinder in Canada, and they tend to co-operate more. In the United States I think their first inclination is to beat each other up." But Rogers makes no bones about his concerns— the weak dollar may be an advantage in some respects, but it wreaks havoc on a business that depends on imported components. He also maintains that there is a danger, both at an individual level and at a national level, of resting on our laurels.

Like many other "success" stories in this book, Rogers has also had his ups and downs. He likes to tell of his early days, "When we just didn't have sufficient cash to pay all the suppliers, I put all the bills in a hat. After meeting the payroll, I'd keep drawing the invoices until we ran out of money. When the creditors would yell at me and I asked them to stop, they sometimes angrily asked what I would do if they kept yelling. 'Very simple,' I said. 'I won't put your invoice in the hat next week.'"[1]

---

[1] "Ted Rogers: How to Build a Business," *Growing Your Business, Profit Guide*, February 2001.

INTERPRETATION VS. INVENTION

INNOVATOR_ ELLIE RUBIN, SERIAL ENTREPRENEUR, COACH INNOVATION_ FOUNDED ONE OF THE PIONEER FIRMS IN DIGITAL ASSET MANAGEMENT SOFTWARE WHICH WAS SOLD TO DOCUMENTUM

Many people believe that to be an entrepreneur requires an ability to invent something new. Sometimes that's true, but Ellie Rubin reminds us that in today's fast-paced, competitive world it may be even more important for those who "entrepreneur" to recognise trends, to read signals, and to create a model of business based on one's ability to *interpret*, not invent. As co-founder of the Bulldog Group, a full-service marketing and design company that grew into the world's leading creator of digital asset management software, Rubin tells her own entrepreneurial story as a prime example of what it takes to become a master of interpretation.

Rubin and her business partner, Christopher Strachan, founded the Bulldog Group in 1991 to design and develop graphics on a fee-for-service basis with a focus on multimedia design, specializing in CD-ROM production and website development. But all along they intended to leverage the knowledge they gained from providing digital and design services for Fortune 500 clients into something much broader. In the course of creating content, Rubin and her partners gained insight into an emerging problem: organizing, retrieving, and re-purposing the masses of audio, graphic, and video files. It was an enormous challenge both within the studio and within their clients' organizations. Rubin saw, early in the game, that the explosion in multimedia content would bring a huge demand for managing content. But Bulldog was way, way ahead of the curve. "People didn't know what we were talking about. Media management? In the mid-nineties no one had a clue what that term meant." Nevertheless, Bulldog took a leap and developed digital asset management software, or "middleware," that was designed to store and index rich, complex digitized content—graphics, audio, and video—so that it could be easily retrieved, re-purposed, and distrib-

> "PEOPLE DIDN'T KNOW WHAT WE WERE TALKING ABOUT. MEDIA MANAGE-
> MENT? IN THE MID-NINETIES NO ONE HAD A CLUE WHAT THAT TERM MEANT."

uted. The software met a specific but critical need. Studies have since shown that people in broadcasting, entertainment, and large corporate organizations may spend more than 10 percent of their time fruitlessly searching for in-house media files. Rubin says the resulting innovation came from "matching interpretation with an enterprising attitude." For example, she took the already acquired knowledge and understanding of how people produce media, and instead of coming up with something entirely new, she and her partners built a business model that mapped their interpretation of market needs with a pragmatic and elegant solution to manage media files. In essence, they were creating a way to leverage what is most valuable to companies—their highly coveted intellectual property.

Now that Rubin was convinced that Bulldog was in the software business, it was time to compete in the software mecca, Silicon Valley. So, in 1994, she packed up her car

---

AS AN ENTREPRENEUR, RUBIN ALSO MADE A COMMITMENT TO TAKE RISKS ALONG THE WAY. SHE USES ONE OF HER FAVOURITE LINES FROM KURT VONNEGUT TO SUM UP THE PRACTICE: "DO SOMETHING EVERY DAY THAT SCARES YOU." IT'S A RITUAL THAT RUBIN INSISTS WILL CULTIVATE GREATER CONFIDENCE FOR EMBRACING UNCERTAINTY—WHICH IS THE ONE CERTAINTY WE CAN ALL DEPEND ON.

---

and left home (against the advice of family and colleagues) to open shop in the U.S. Rubin wasn't an engineer or a computer scientist, but she did have "calculated intuition" and she knew that they had the right concept at the right time. "And no matter how many people told me I couldn't do it and I shouldn't do it, the more I knew I was going to do it," she says. As an entrepreneur, Rubin also made a commitment to take risks along the way. She uses one of her favourite lines from Kurt Vonnegut to sum up the practice: "Do something every day that scares you." It's a ritual that Rubin insists will cultivate greater confidence for embracing uncertainty—which is the one certainty we can all depend on.

This time the risk paid off. The niche that Bulldog had envisioned before anyone else expanded far beyond the entertainment sector that Bulldog had initially targeted. Bulldog's software sold internationally to broadcasting, entertainment, publishing, and Fortune 500 marketers. In addition to the investment of BCE and other Canadian investors, convergence giant Sony Pictures Entertainment and technology player Sun Microsystems became clients, then investors in the company. Other clients include Microsoft Studios, Sears, EMI, and the BBC. Most recently, Bulldog was bought by Documentum and continues to sell the digital asset management product internationally. According to market researchers Frost and Sullivan, the asset management market is one of the fastest-growing segments in the industry, expected to reach $2.4 billion by 2006. Looking back, Rubin says it took an ability

FOR ELLIE RUBIN, "ENTREPRENEURING" HAS ALWAYS BEEN ABOUT EMBRACING RISK AND IMBALANCE, MASTERING INTERPRETATION RATHER THAN INVENTION, AND TURNING EVERY CHALLENGE INTO AN ADVANTAGE.

to identify and interpret trends that gave her and her partners the courage to move the company from a conventional service-based industry to multimedia production and finally to becoming a software product development firm. In the end, those radical shifts were critical for Bulldog's success and perhaps even necessary for its survival.

Rubin also has some keen observances about Canadian competitive advantages. By doing business in the "idiosyncratic landscape" of Los Angeles and San Francisco, she discovered a handy business tactic, the chameleon-like ability to fit comfortably into any business environment. "As Canadians, we know how to utilize the tactic and do it well. By adopting the image of the chameleon, we could walk into any meeting, any boardroom, and take on an unassuming pose. We often play the highly researched, well-prepared, but relatively low-key, player at the table." Rubin believes business chameleons are able to blend in and take the time to interpret nuances within any given organization. When dealing with entertainment and film studios or large production houses, this ability to tap into the particular culture of a group of decision-makers is crucial to getting the deal closed. (There is a warning here. This chameleon-like approach must be balanced with a strong promotional backbone, so that an entrepreneurial company's ability to brand itself is never lost in the process of gaining new customers and competing effectively in an overcrowded market. It is a fine and delicate balance to maintain, but one that is powerful in times of change.)

Never one to rest on her laurels, Rubin has recently launched a new company, Ellie Corporation. Once again as a master of interpretation—not invention—Ellie leveraged the accumulated knowledge she gained in moving to the Silicon Valley, raising investment and launching a design and software company. This time around she has carved out another niche for herself as an international speaker, author, columnist, and television host. This career phase was kick-started by the publication of her best-selling book, *Bulldog: Spirit of the New Entrepreneur*. With the launch of her latest enterprise, Rubin continues to take her own advice. "Entrepreneuring" is not just for those who want to start their own businesses—it's a mindset for success. Rubin uses the term as a "relevant concept for anyone who wants to create their own 'story'—whether that means starting one's own company, managing someone else's, or simply trying to incorporate a spirit of entrepreneurship into the work one does." For Ellie Rubin, "entrepreneuring" has always been about embracing risk and imbalance, mastering interpretation rather than invention, and turning every challenge into an advantage. So what's her advice for how to start out on the "entrepreneuring" path? "Take what you've learned and accumulated during your lifetime and then jump off a new cliff."

SHARING THE PODIUM

INNOVATOR_ JONATHAN SEELIG, SERIAL ENTREPRENEUR INNOVATION_ CO-FOUNDED AKAMAI, A WORLD LEADER IN E-BUSINESS INFRASTRUCTURE AND CONTENT DELIVERY

Often entrepreneurs start their own companies because they want to control their own destinies. Their drive, singleminded commitment to "the vision," and supreme confidence are often factors contributing to their success. But such magnificent obsessions can also lead to failure when entrepreneurs become unable or unwilling to relinquish control over fast-growing and increasingly complex organizations. Jonathan Seelig, co-founder of Akamai Technologies, believes strongly that sometimes the most successful entrepreneurs are those who are willing and able to hand over control of the day-to-day operations of their businesses to experienced managers who bring other skills to the table.

Vancouver-born Seelig was an MBA student at MIT's Sloan School when the promise of adventure and fortune led him to abandon the classroom for "the real world." In his first semester he met up with professor Tom Leighton and graduate student Daniel Lewin, who were working in MIT's lab for computer science. They were trying to respond to a challenge by Tim Berners-Lee, the inventor of the World Wide Web, and develop mathematical algorithms that could help reduce Web congestion and more efficiently deliver content to Web users. They reasoned that by placing the heaviest loads near the user they could eliminate bottlenecks and optimize transmission speeds. The trio began discussing ways to commercialize their technology and in 1998 they entered MIT's esteemed annual "$50K Entrepreneurship Competition," where the company's business proposition was selected as one of six finalists among 100 entries. In 1999 Akamai launched its first services with an IPO later in the year. The name means "intelligent, clever, and cool" in Hawaiian. It is now synonymous in the industry with "really big, really fast." Within two years the company was worth a staggering $20 billion (U.S.).

Timing is everything. Akamai essentially created a market, dubbed "Content Delivery Alternatives" (CDA). Instead of developing its own infrastructure to compete with ISP giants, the company elected to work with established Internet service providers across the country to provide a value-added service. In fact, the company's goal has been described as "to manage the largest cohesive, managed IP network in the world—without actually owning the networks they're on." Today, the company operates more than 13,000 caching servers, which sit on the edge of more than 1,000 networks in 63 countries to distribute content and applications. These servers constantly analyse traffic flow and re-route information away from potential congestion, essentially acting as an extension of an organization's internal network to dramatically increase speed and improve performance. Akamai can speed the delivery of streaming video, PDF files, and other forms of content. The company has developed a solution aimed at addressing the slow speeds of "the last mile" that typically links high-speed corporate networks to interorganizational carriers. Customers include companies like Apple, FedEx, Motorola, Time Warner, and Yahoo.

Although the carriers are nipping at their heels with stripped-down versions, Seelig says the company will stay ahead of others by adding applications that are too complex for others to launch. Technological development remains a priority for sustaining their market position, and the company is obsessed with enhancing security and reliability. Seelig's role is to develop new technology and business initiatives to support Akamai's core business and growth. He knows what he knows, but also what he does not know.

He and his partners knew from the start that their vision depended on getting very big, very fast, so they would have something that is not easily duplicated. The challenge of growing from three people to several hundred in a short time was phenomenal. "I'm learning what they can't really teach you in school. It's really an awesome job," he

> **"YOU NEED TO KNOW NOT JUST WHAT YOU'RE GOOD AT, BUT WHAT THE GAPS ARE AND THE SKILLS THAT YOU NEED TO FILL. RECOGNISING THE VALUE OF THOSE OTHER SKILLS AND ASSETS, THE ONES YOU DON'T HAVE, IS IMPORTANT. UNDERSTAND WHICH ONES REALLY MATTER TO YOU AND WHERE YOU'RE GOING TO RECRUIT THEM FROM. YOU NEED TO HAVE A PLAN OF WHO'S GOING TO HELP YOU FILL THOSE THINGS."**

says. Strategic and technological brilliance aside, Seelig is very clear about the basis of this remarkable success—learning how to work closely with those who had both the know-how and resources needed to move the enterprise to the next level.

Some entrepreneurs are wary of venture capitalists, concerned that VCs will want to seize control of the business and squeeze out the founders. But Akamai embraced them. Before long they were working closely with Battery Ventures, a Boston-based venture capital firm, to develop a business model to help "three guys trying to figure out what to do with this math that Tom and Danny had developed. Our experience with our VCs was terrific. I would say wholeheartedly it was very, very good. They spent a good nine months with us before funding us, working through the kinks in the plan, and really making sure that there was, in fact, a business here."

Unlike many who are blinded by their own brilliance, Seelig and his co-founders understood that they couldn't handle all aspects of the new venture on their own. "You need to know not just what you're good at, but what the gaps are and the skills that you need to fill. Recognising the value of those other skills and assets, the ones you don't have, is important. Understand which ones really matter to you and where you're going to recruit them from. You need to have a plan of who's going to help you fill those things." They have also been very good at developing relationships to strengthen their position, a key element of their strategy. In addition to working closely with ISPs, Akamai has partnerships with industry players like Oracle, IBM, Sony, and Ericsson.

What many entrepreneurs forget is that there are lots of talented people who can help them develop their businesses. And good advisors will help entrepreneurs find that talent. "We found that we were very successful in building some of the skill sets that we didn't have through our investors. Our venture capital investors helped us specifically with some work on international expansion into Europe, for example. They helped us to recruit terrific people, including our president and our chief executive officer." Paul Sagan, Akamai's president, joined the company in 1998. CEO George Conrades joined in 1999. Leighton is Akamai's chief scientist.

With the help of their VCs, the Akamai team found the right people to help them manage the company. "We brought in people very early on to help us grow the business significantly. We had a CEO come in who respected our place, our opinion, and our ability to contribute substantially to the direction of the company. The company didn't need to be his; it could be all of ours together. That was very healthy for us."

Not all entrepreneurs are so open to involving other people in their companies, particularly at the early stages. But in Seelig's opinion, this ability is a characteristic that defines successful entrepreneurs. "The folks I've seen who have been successful, who've germinated good ideas and then been successful in turning them into great companies, have been people who also know when to step aside and bring in more experienced leadership, or bring in people with skill sets that they don't necessarily have."

The Akamai experience shows that Seelig and his partners understood this critical success factor. But in Seelig's opinion, "the great example of this is Pierre Omidyar, at eBay. He's one of the most successful entrepreneurs that this age has seen, and he knew when to turn over the reins. He also recruited people who were very good at allowing him to continue to control the vision and the idea of community at eBay for a long time. And when you hear Meg Whitman, eBay's CEO, talk about him, she talks about the fact that he also knew when to let her run the business, because she is a great manager."

In the end, it appears that the real challenge is finding the right balance of control within a company. Entrepreneurs want to be involved in setting the strategic directions for their companies. As Seelig says, "You need to be the biggest believer in what you're doing. When you find the day that you're no longer the biggest advocate and the biggest believer, you have to evaluate whether it's still your company and whether you should still be there." If the answer is no, the entrepreneur exits, and begins thinking about the next great idea.

Seelig's philosophy and principles have defined an entirely new market and built a huge, world-class company in a few short years. Life, Seelig learns, can throw unimaginable obstacles in your way. The whole industry reeled in shock when his friend and co-founder Daniel Lewin, the company's chief technology officer, was tragically killed in the attack on the World Trade Centre. Taking time to reflect, Seelig left Akamai in the spring of 2002 to focus on new ventures. The two have, however, left a legacy and a solid foundation that should serve their successors well.

POETRY AND CODING

INNOVATOR_ GERRI SINCLAIR, SERIAL ENTREPRENEUR, ACADEMIC BRIDGE INNOVATION_
FOUNDED INDUSTRY-LEADING CONTENT MANAGEMENT FIRM NCOMPASS AND SOLD IT TO
MICROSOFT

"Okay, just give me a half-time salary and room to grow, and I'll attract the funding that will allow us to put Simon Fraser on the map as a place of innovation!" It was 1987 and if you happened to be walking by the office of the dean of education at SFU in Burnaby, British Columbia, you might have overheard Dr. Gerri Sinclair getting the go-ahead to establish the first entrepreneurial group of its kind—aimed at creating and selling leading-edge interactive educational technology. Much has been written on the importance of talent, research, and innovation to Canada's leadership in the digital economy. But getting the ideas out of the research labs and into the marketplace is often a difficult process. The skills that bring success in academia are almost antithetical to those associated with success in business. Sinclair is an exception.

Hers was a revolutionary request, since "entrepreneur" and "educator" had seldom been said in the same sentence. In fact, there was no existing formula for transforming an R&D model into a revenue-generating business, but Sinclair didn't let this hold her back. With the blessing of her forward-thinking dean, she founded ExCITE (The Exemplary Centre for Interactive Technologies and Education), the first multimedia research and development centre in Canada. Sinclair was its only employee. However, within a short time, the group had grown to 35 people and the research team had not only become specialists in creating Internet software, but also successfully sold their products and services into the marketplace. ExCITE pioneered many multimedia initiatives, including *Science, eh?* (the world's first multimedia science magazine); "*InfoProbe*," an interactive Internet kiosk that introduced the public to the World Wide Web; and "*The Prime Ministers of Canada*," an ambitious interactive exploration of the lives and times of Canada's first Prime Ministers. ExCITE also developed the first website in Canada, using Mosaic, the precursor to Netscape, as a browser.

Sinclair was on a roll. Soon she teamed up with a couple of talented programmers from the lab and cooked up a technology that immediately caught the attention of software giant Microsoft. It was 1995, the height of the Internet browser wars, and ExCITE's researchers had devised a plug-in that enabled Microsoft's Web-development tools and Office applications to run on Netscape's browser. Soon, word reached the outside world about this innovative browser solution, and the ExCITE crew was inundated with requests from wannabe partners, investors, and even some talent poachers.

It was a heady time for the university-based team, and Sinclair decided she was set to move from her safety net of the nonprofit realm to face the challenge of succeeding "in the 'real' world." She and a handful of researchers left the lab to forge their own private company, NCompass. Moving from an R&D culture to one based on commercialisation was a daunting challenge for the former English professor. Sinclair admits she had her hands full coping with common entrepreneurial tasks, such as "scaling and growing the business, and hiring and retaining talent, while attempting to secure startup capital." Right from the get-go there were doubts raised about her abilities to compete in the corporate environment. Some even questioned Sinclair's decision to act as CEO of her own company, including one potential investor who said, "You can't seriously be thinking of being the CEO of this company." Sinclair replied, "Now, why would that be?" and he said, "Well, first of all you have no business experience, and second of all, you're a woman." That was 1995; Sinclair says comments like that convinced her to take the helm against all odds.

MUCH HAS BEEN WRITTEN ON THE IMPORTANCE OF TALENT, RESEARCH, AND INNOVATION TO CANADA'S LEADERSHIP IN THE DIGITAL ECONOMY. BUT GETTING THE IDEAS OUT OF THE RESEARCH LABS AND INTO THE MARKETPLACE IS OFTEN A DIFFICULT PROCESS. THE SKILLS THAT BRING SUCCESS IN ACADEMIA ARE ALMOST ANTITHETICAL TO THOSE ASSOCIATED WITH SUCCESS IN BUSINESS. SINCLAIR IS AN EXCEPTION.

Nevertheless, Sinclair's recipe for leadership was based on a mixture that even she would admit consisted of one part innovation and chutzpah and two parts ignorance and inexperience. With nary a business plan in sight, or, for that matter, any real revenue-generating model, Sinclair concedes she broke all the rules. Now faced with the task of pitching potential investors rather than university deans, her approach was equally unorthodox. (For those of you seeking venture capital, you may not want to use Sinclair's approach as a model.) Her spiel ran something like this: "I can make up a business plan and tell you we're going to sell X number of these over X number of years, but I don't believe it. And I'm not going to do that. If you're going to invest in this venture, it's because you believe that this group has the technological prowess and entrepreneurial spirit that can actually be successful." Honesty turned out to be the best policy. NCompass gained high-profile backers like U.S.-based computer game maker Electronic Arts and chip giant Intel. Put that in your MBA and smoke it.

But how did this former English professor make her own transition from educator to technology entrepreneur? The draw for Sinclair, who has a Ph.D. in Shakespearean drama, was the "transformative power of technology especially for writers, artists, and musicians." She says, "I soon got hooked on programming, because I looked at coding much the same

> SHE ADMITS EACH EXPERIENCE REQUIRES A DIFFERENT PERSPECTIVE, BUT SHE KNOWS THAT "BREAKTHROUGH THINKING" IS MANDATORY FOR SUCCESS AND SURVIVAL. SHE ALSO KNOWS THAT A GOOD IDEA MEANS VERY LITTLE WITHOUT EXCELLENT EXECUTION.

way as I looked at poetic analysis. You compare the structure of a work of art to a computer program and they are not all that different. And for me, code became the new medium of creation that changed my life."

As a writer herself, Sinclair recognised that content was key, and this knowledge later gave her company a competitive edge. "It was obvious to me that there was going to be exponential growth in the amount of information in various formats that would be published on the Web, and that most corporations would need tools in order to help manage that information, to allow potentially everyone in the organization to become their own publisher." Creating content-management solutions built on the Microsoft platform became NCompass's "hot" business case and led to the recent opportunity to sell the company to

> "IT WAS OBVIOUS TO ME THAT THERE WAS GOING TO BE EXPONENTIAL GROWTH IN THE AMOUNT OF INFORMATION IN VARIOUS FORMATS THAT WOULD BE PUBLISHED ON THE WEB, AND THAT MOST CORPORATIONS WOULD NEED TOOLS IN ORDER TO HELP MANAGE THAT INFORMATION, TO ALLOW POTENTIALLY EVERYONE IN THE ORGANIZATION TO BECOME THEIR OWN PUBLISHER."

the computer giant to the tune of $55 million in cash. It was a quick and enticing deal, yet the decision to sell was tough for CEO Sinclair, who was "so close to realizing a dream of building a world-class Canadian software company." By then, NCompass had grown to 180 people and was exceeding all revenue targets, so although she was delighted by the success, in some ways Sinclair says she found the sale heartbreaking. Nevertheless, Sinclair says Microsoft's tremendous marketing power will give the product the global reach NCompass might never have achieved on its own.

Today Sinclair is president of the Premier's Technology Council, a group of technology leaders and visionaries within the British Columbia technology community, advising the Premier on technology-related matters in the province. Specifically, Sinclair is focused on building a robust B.C.-based technology industry and on bridging the digital divide by bringing affordable broadband access to all communities in the province. With this new appointment, her pioneering work with technology has now intersected three realms—academia, the private sector, and now the public sector. She admits each experience requires a different perspective, but she knows that "breakthrough thinking" is mandatory for success and survival. She also knows that a good idea means very little without excellent execution. Renowned for her ability to get the steady stream of good ideas out of the labs and to the customer, Sinclair is ready to tackle the next round of technological challenges facing Canada this century.

DEAL DIVA

INNOVATOR_ SUKHINDER SINGH, SERIAL ENTREPRENEUR, FINANCIER INNOVATION_ HELPED
BUILD ACCOUNT AGGREGATION SOFTWARE LEADER YODLEE.COM AND RAISED $85 MILLION (U.S.)
IN THE PROCESS

Great ideas are a dime a dozen in Silicon Valley. Turning great ideas into successful business ventures requires much more, including the ability to attract and retain capital and customers. Many startups don't make it because they fail to inspire the confidence and support of investors and major accounts needed to play in the big leagues. Although Sukhinder Singh may not look like a stereotypical power broker, the 31-year-old from St. Catharines, Ontario, is turning heads. *Business Week* has singled her out as a successful dealmaker.[1] Singh helped secure $85 million in start-up financing for Yodlee.com in a little over two years. Support has come from private, financial, and corporate investors and

---

[1] Constance Loizos, "The Rise of a Dealmaker," *BusinessWeek* online, February 27, 2001, edited by Jennifer Gill.
http://www.businessweek.com/careers/content/feb2001/ca20010227_960.htm

advisors, including the founders of Hotmail, Junglee, Exodus, and Integrated Systems, as well as executives from Amazon.com, Netscape, Microsoft, Checkpoint Software, AT&T, AOL, and VXtreme. Bob Sandler, director of Channel Strategy at AOL, which has been an active user of Yodlee's technology, says that Singh is "as good as anyone I've ever worked with," and adds, "I still remember her proposals on how our deals might be structured."[2]

Singh is also credited with helping Yodlee make remarkable in-roads in the financial industry, which tends to be particularly risk adverse compared to other high-tech investment sources. Already Yodlee has signed 60 percent of the top 25 banks and about 80 percent of the top U.S. brokerage companies including Merrill Lynch, Bank of America, and Morgan Stanley Dean Witter. Yodlee's technology enables financial institutions and portals to pull together information on their customers' bank accounts, credit card balances, frequent flyer plans, stocks, and even their e-mail habits. Singh says that Yodlee's technology "enables financial clients, for the first time in history, to create a holistic picture of their customers and then market financial services directly based on their specific needs." It is because of this unique aggregating feature that Singh asserts Yodlee's technology has gone from "nice to have" to "must have" for the world's financial sector. She has played a critical role in helping them understand that need.

So, how did Singh get from St. Catharines, Ontario (population 130,000) to being the darling of the U.S. business press? Born to Indian parents, she came to Canada when the family emigrated from East Africa in 1975. Singh eventually went on to study business administration at Western University's Ivey School and she credits the rigorous case-driven program for cultivating business excellence and training her to think on her feet. After graduating, she joined Merrill Lynch in New York. There we see the first signs of her skills—in her first three months she managed an IPO that raised $300 million. Her former boss Henry Michaels says "Almost from day one Sukhinder was operating as the functional equivalent of a senior associate."[3] She moved with Merrill Lynch to London, England, and then to a new job working for Sky Broadcasting. But with her career as an

[2] Ibid.
[3] Ibid.

executive at BskyB in full flight (working for the CEO and COO on strategic initiatives), Singh walked away from a recent promotion and six-figure salary. She wanted a new challenge and she wanted to come back to North America. So after a quick trip home, she bought a one-way trip to L.A., bought a car with her remaining cash, and made a virtually penniless trek to northern California to tap into the technology sector. "A lot of what I do is on impulse," she says. "I felt like I needed a change." She ended up with some of her university

> **TURNING GREAT IDEAS INTO SUCCESSFUL BUSINESS VENTURES REQUIRES MUCH MORE, INCLUDING THE ABILITY TO ATTRACT AND RETAIN CAPITAL AND CUSTOMERS. MANY STARTUPS DON'T MAKE IT BECAUSE THEY FAIL TO INSPIRE THE CONFIDENCE AND SUPPORT OF INVESTORS AND MAJOR ACCOUNTS NEEDED TO PLAY IN THE BIG LEAGUES.**

friends in northern California, where she came across a startup called Junglee. Although she thought the technology was boring, she was impressed with the founders and agreed to help them build some relationships with e-tailers that they needed for a deal with Yahoo. Singh knew how to write business plans. She also knew how the financial sector operated. Junglee's business grew by leaps and bounds and in August 1998, Amazon.com acquired 1.6 million shares for $190 million and invited Singh to come on as a business development manager. At Amazon.com, Singh developed an attractive revenue-sharing model to bring on retailers. From there, Singh was actively courted by the Yodlee team, led by Venkat Rangan.[4]

---

4 Ibid.

Joining Yodlee (the name, like "Junglee," is derived from an old Indian song) also united Singh with a community of Indian colleagues and culture, a network she hadn't forged ties with previously. Like many Canadians, she straddles two cultures. Ironically, her interest in new technology ended up connecting the "old world" of her family.

Now considered a veteran of the tech industry, Singh has impressed co-workers, clients, and even competitors with her passion, integrity, and style. "She goes well beyond winning by ensuring that she does right by the partner and frequently represents the customer's point of view to help us better products for Yodlee's market," praises Ram Shriram. Even those who have faced Singh from the other side of the negotiating table give her high marks for keeping it clean. Tim Swords, a VP with Fidelity Brokerage Company, admits, "She's tough, but ultimately reasonable and professional." His company came onboard as a Yodlee partner over two years ago and, thanks to Singh, has become one of their biggest customers. By all accounts she inspires the confidence and respect needed to get others to sign on the bottom line.

Despite her remarkable successes, Singh confesses that the rapidly changing landscape of high tech presents a constant challenge. Although she had played hardball in the investment world, Singh says there was a time when making deals with companies many times the size of Yodlee made her very nervous. The trick, she says: "Don't say 'yes' just because you're small. You *can* say 'no' to a big company." She laughs and admits, "They don't like it and they threaten, but more often than not they'll come back to the table."

Singh soon overcame the size hurdle and now confidently shares a wealth of other dealmaking advice. Keeping pace with "timing" in the turbulent technology space can mean everything, Singh warns. She recalls with frustration an instance when Yodlee was trying to close a deal with a major online player. After seven months of exhaustive rounds of negotiation, Yodlee was ready to put pen to paper, only to realize that the deal terms were no longer appropriate. In the time it took to reach a consensus, Yodlee's market position had changed and the general market had fluctuated, making the agreement they were about to sign obsolete. Now, Singh attempts to instill a "sense of shared urgency" between all parties in order to quicken the pace. That's not the only reason Singh cites for keeping negotiation cycles short. "The sooner you cut a deal, the easier it is to avert attempts by the competition to bungle your deal," Singh states. She remembers another time when Yodlee was a mere 24 hours away from signing a $500,000 contract with a major financial institution. As negotiations laboured, a competitor camped out on the potential client's doorstep and managed to create doubt in the firm's mind as to Yodlee's capability. They never did close that deal.

> **NOW CONSIDERED A VETERAN OF THE TECH INDUSTRY, SINGH HAS IMPRESSED CO-WORKERS, CLIENTS, AND EVEN COMPETITORS WITH HER PASSION, INTEGRITY, AND STYLE.**

Through dealmaking lessons learned the hard way, however, the Silicon Valley-based company has not only survived, it has flourished. Now with offices in Atlanta, Bangalore, and London, Yodlee has gone global. "We've expanded not only domestically, but internationally," says Singh, and she confidently predicts that Yodlee will be profitable by early 2003. Yodlee also has its sights set on Canada. Having conquered the financial sector in the U.S., Singh says that it makes sense to come back home and sell their account aggregation software to the less fragmented Canadian banking sector.

What will happen next? It's anyone's guess, but by any measure Singh's success has been remarkable: Bud Colligan of Accel Partners, which has $10 million (U.S.) invested in the company, calculates the rate of deals made since the company started (about 110 in the first 400 days). "It breaks down to a deal about every three days, work that is "all Sukhinder. She is an incredibly prolific dealmaker. She is always going. Always."[5]

⁵ Ibid.

Tony Davis Mark Skapinker_ Rick Nathan_

ZIGGING AND ZAGGING

INNOVATOR_ MARK SKAPINKER, RICK NATHAN, AND TONY DAVIS, SERIAL ENTREPRENEURS, FINANCIERS INNOVATION_ DEVELOPED MANY WORLD-RENOWNED SOFTWARE FIRMS, SUCH AS DELRINA, BACKWEB, AND BALISOFT

Mark Skapinker, Rick Nathan, and Tony Davis, founders of the Brightspark Group, a company that supports up-and-coming software developers, have turned "zigging and zagging" into a successful management approach. It is not a concept found in any management book. It is about recognising that there are many ways to reach a goal rather than stopping just because the chosen path appears blocked. When they meet an obstacle, they don't stop. They simply rethink their approach to find a way around it.

Success breeds success. Skapinker and Davis's first big win was at Delrina, where Davis developed the popular WinFax software. When they sold Delrina to Symantec in 1995, each then started new companies in the technology sector. In partnership with lawyer Rick Nathan, they formed Toronto-based Brightspark Group using their innovative approach to problem-solving to build and nurture a new generation of Canadian software companies. After June 2002, Nathan moved on to lead Toronto-based Goodman Ventures.

It may seem obvious that successful products are ones that solve real problems for real people. But as often as not the initial solutions entrepreneurs devise to tackle problems may be off-target or unworkable. And when the product fails, many aspiring entrepreneurs just change direction or simply give up, leaving the problem unsolved. But the Brightspark Group has shown that in the current environment of rapid and unpredictable change, persistence and a continued focus on solving problems can yield success where others have failed.

They offer many examples of their approach in action. Skapinker initially started Balisoft to help companies provide support and service to customers purchasing on the Web. Balisoft's LiveContact was developed to allow customers to use Voice over IP (VoIP) to speak directly to a customer service representative while they were browsing the company's website. Not long after LiveContact was launched, "Voice over IP collapsed. It wasn't happening. The infrastructure didn't work." So the VoIP product was abandoned, but not the effort to find a solution to the problem. Balisoft went back to its customers, who "were telling us they're using e-mail for customer service. 'You can't believe how many e-mails we're getting,' the customers said." Skapinker realized that the problem of how companies could provide customer support over the Internet hadn't disappeared, but the solution lay in a different technology. "We turned the company around. We dropped LiveContact, and we started up a new product called EmailContact." The lesson? The focus remained on solving the problem, not on the technology per se. The question was "How do you answer lots of e-mails, and how do you become more responsive?" As long as you're focused on your end game, never mind how many times you zig and zag, you've got to get from here to there.

Skapinker offers another example of a company that managed to get "from here to there." Borderfree Inc., one of Brightspark's portfolio companies, "had this great consumer product that could let you buy off any U.S. Web page." The company knew that "e-commerce is great until you're actually going to buy something. 'Welcome to this huge mall. Welcome to these great stores. Welcome to all this stuff. If you want to buy it ... not so fast. You're

---

ZIGGING AND ZAGGING IS NOT JUST AN APPROACH; IT'S A MINDSET. "TO BE SUCCESSFUL IN THIS INDUSTRY, YOU HAVE TO REALLY LIKE CHANGE; YOU HAVE TO EMBRACE IT. YOU CANNOT BE SOMEBODY WHO HAS A PLAN AND JUST WANTS TO FOLLOW IT NO MATTER WHAT. SUCCESSFUL COMPANIES, PARTICULARLY AT THE BEGINNING, ARE CHANGING THEIR BUSINESS PLAN EVERY COUPLE OF MONTHS, SOMETIMES RADICALLY. "

Canadian; you can't buy it. We don't take your credit card. We don't recognise your address. We don't know how to ship it to you. Don't know how to get it to you. We don't know how to deal with you.'"

Initially the idea was to make it easy for Canadians to complete online transactions with U.S.-based companies. Canadians would go to Borderfree's website to place their orders with U.S. retailers, and Borderfree would handle getting the goods across the border. Good idea? Maybe, but the dot-com meltdown meant that Borderfree couldn't raise the capital it needed to tell Canadian consumers what a great service the company was offering. Faced with a lack of funding and low consumer demand, many dot-com companies simply shuttered their websites, and gave up on e-commerce. But once again, Borderfree did not abandon its initial quest. It just took a slightly different approach to making it easy for Canadians to buy things on the Web. The company realized that if it couldn't run its own website, another way to improve the purchasing process was to make changes on the vendor site, and this is what they did. "Companies have to add five lines of code to their entire Web page—and they can switch on Canada. We worry about verifying the credit card. We worry about getting it to them. We worry about everything. It's built directly into a company's products and they pay us a piece of their transactions to Canada," Skapinker says. Borderfree has opened doors, allowing companies to expand into Canada with little upfront investment.

Tony Davis tells a story about developing applications for use on mobile phones. After three days of work, an engineer came to Davis, pleased with an application he had built. What was Davis's reaction? "I looked at the application and said, that's phenomenal. But what you've just taught me is that anyone can create these things in three days." What did this mean for the development project? Since there were very few barriers to entry, Davis decided he didn't want to be in this business. But rather than abandon the idea, he "zigged and zagged" to find an alternative way to solve the same problem, one that was much more difficult to duplicate.

Zigging and zagging is not just an approach; it's a mindset. "To be successful in this industry, you have to really like change; you have to embrace it. You cannot be somebody who has a plan and just wants to follow it no matter what. Successful companies, particularly at the beginning, are changing their business plan every couple of months, sometimes radically. We have seen individuals fail where they didn't have adaptability. It's not just a matter of being able to roll with the punches, you have to actively like it. You have to want to go into an environment where the ground is moving under your feet all the time. Some people love it and thrive in it, and some people hate it."

The Delrina partners were all immigrants, and Skapinker thinks this explains some of their drive to succeed. "We grew up in Southern Africa. And we'd arrived here and decided, this is where we were going to live, this is where our kids were going to grow up, this was the place that we really wanted to be. And that we really wanted to be successful, and successful fast. There was a complete drive—no matter what happened—to be successful at it. I think that there's an immigrant mentality attached to that drive. No matter what happens, I'm going to succeed. There isn't any infrastructure around me to fall back on and it's not my family, or my dad, or my anybody else that is going to bail me out; it's me doing this and I have to succeed no matter what, and I'm on my own with this. And that was always an underlying part of it."

It's that unswerving dedication to success that really differentiates successful entrepreneurs from also-rans. "The non-entrepreneurial thinkers keep saying, 'Well, I spoke to all of these experts in the area and they told me that if we want to get into this area, it's really difficult. You really need experts to help you, you can't do it on your own.' The entrepreneurial thinkers say, 'Well, those people did it all wrong, that's why they failed. I can do it because I'm not going to allow myself to fail. I'm sure that if I apply the principles that I've got toward how we should do things, then we can then do it and completely differently, and we can pull it off.'" As Skapinker says, "In the face of everybody telling you what you can't do, just determining what you can do makes the difference. When we started building Delrina, it was February 1988. It was three months after the crash of '87; that's how I remember it so well. And everybody said to us, 'You're crazy. You've never run a software company before. The stock market has crashed. You're sitting in Canada. You're not from here. What are you doing? Why don't you just get a real job and stay focused on earning a living instead of doing this now at the worst possible time? It's the worst this, it's the worst that. Nobody's every done this, you're never going to pull this off. Forget it, don't do this.' And we just said, 'Well, no, we see an opportunity and we can do it. We've got enough confidence that we'll make this successful no matter what, and zig and zag our way through whatever hurdles that we get to, and we'll figure out how to do it.'" And the partners' sequences of success have showed that they are crazy indeed. Like a fox.

THE "SOFT" SKILLS ARE HARD

INNOVATOR_ CAROL STEPHENSON, INFRASTRUCTURE BUILDER INNOVATION_ KEY ARCHITECT BEHIND THE STENTOR ALLIANCE, TURNING STATE-RUN MONOPOLIES INTO A WORLD-CLASS COMPETITIVE COALITION

The notion that a strong telecommunications infrastructure is a key to national competitiveness is now almost a cliché, but building that national infrastructure has been anything but easy—particularly given the dramatic changes in the telecom landscape. Companies have had to reinvent themselves, not just in response to technological changes, but also to fundamental shifts in the very structure of their business. Before being snapped up to head Lucent Canada, a subsidiary of the largest telecommunications equipment maker in the world, Carol Stephenson had made a name for herself as a change agent in the telecom sector. Canadian Women in Communications named her Woman of the Year, and *Maclean's* had singled her out as one of the country's top female CEOs. Stephenson had led the formation of Stentor Resource Centre Inc., an alliance among incumbent telephone companies developed in 1993 in order to enable them to compete more effectively. Bringing together the range of telephone companies large and small—from behemoth Bell Canada to tiny Island Tel and NorthwestTel—was no easy feat.

And in a position where progress was dependent on building a consensus, Stephenson was remarkably successful in shepherding others through fundamental shifts. One of her riskier moves was to convince the Stentor companies to open their local business to competition, in exchange for CRTC approval to enter new lines of business. She also persuaded the companies to significantly increase their marketing initiatives, creating a national long distance brand to compete vigorously with the new entrants in the competitive long distance market. For several years, Stentor was a force to be reckoned with and provided the incumbent telecom companies with the competitive advantage they needed to adapt and grow in the big leagues. In 1999, when the alliance dissolved, two stronger players emerged—Telus on the one hand, combining Alberta Tel and BC Telecom with the wireless network of Clearnet, and Bell Canada Entreprises on the other. Establishing and implementing a common vision among such diverse players was sometimes like herding cats. Along the way, Stephenson learned the incredible power and influence that can be wielded through what she calls "soft skills."

Stephenson learned the importance of soft skills early in her career. "I was having my performance evaluation. I brought in stacks of books and numbers and everything to prove my worth. And my boss said, 'Just put all those numbers aside; what I would really like to talk to you about is leadership.' I was complaining about some fabulous idea I had that had not been accepted, and he said, 'Well, there are two reasons for that. The first could be that it's a dumb idea; the second could be that you didn't sell your idea very well. And knowing you, it's probably the second, not the first. In other words, you didn't influence.'" That's a lesson that has served her well. "When something doesn't go exactly as I think it should, I do a little introspection and ask, 'How did I go about selling this idea? Could I have made better points? Did I talk to the right people? Did I do enough influencing?'"

Stephenson's leadership rests on her communication abilities. "A lot of people might say you've got to be tough as nails to be a leader. But having soft skills doesn't mean that you do not make hard decisions. I know some people consider consensus a dirty word. Some people say, 'Oh consensus, it takes way too long. You need somebody that can make a snap decision and just make sure it gets done.' I don't think consensus needs to take forever and ever. But I do think that if you can get people rallied behind an idea, or a vision, you are going to be more successful in implementing it. I have always been very good at getting people aligned to a position through a consensus approach." Certainly Stephenson's instincts are bang on in terms of current management thinking—good ideas are only words on paper if executives cannot get the buy-in they need to implement them.

That's where the velvet glove comes in. "It's the communication skills, listening skills, learning about positive reinforcement really early on, that works much better than yelling at people." What's more, she adds, "You can learn engineering, you can learn rules, but the soft skills are ones that you really must work hard at and develop. They're not in a rule book."

Her three rules of management are: "Communicate, communicate, communicate. You can never over-communicate," whether it's with your customers, your employees, or the broader environment. Her commitment to building a healthy corporate culture seems to go beyond the usual lip service. "I just think it makes such huge difference. I've worked in companies where I can see the difference based on the way the leaders lead, and the way you just approach a company. I am always listening. Sometimes they're not big ideas, they're small ideas that can really make a difference, and they all go together to create a culture. Here's a little one we do. We have a thing called the rumour mill, which is online [and anonymous], and we encourage our employees to post rumours and then we answer them. A lot of people think that rumours are bad things. I think rumours are good things because they get out in the open what are the worries and concerns of employees. So why not encourage it? That's a very little idea, but it's just symbolic of some of the things you can do to promote openness."

Broad education and experience are the keys to seeing the big picture. "The broader the experience, the better your ability to synthesise and solve problems. I see so much specialisation going on, not just in companies but also in schools. I'm a firm believer in the need for a very broad general education, and you can specialise afterwards." She also focuses attention on the importance of having the right kind of talent to draw on. Often recruiters say that they need teamwork, organisational skills, and people who understand the application of technology, not just how it works—but those same recruiters still look only for engineers or computer scientists. According to Stephenson, diversity drives innovation. "I'm very clear that I don't want a company entirely of engineers and computer scientists, because they think alike, and you don't get a lot of diversity. They tend to see the world in terms of black and white and there's a lot of grey. I need really good project managers, with strong interpersonal skills, communication skills, leadership skills. And so the ideal is a diverse workforce."

Clearly, in the high-tech sector, scanning the environment is key. "Sometimes you can get so consumed in the day-to-day stuff, you miss the big picture. Don't lose sight of the big ideas." It may seem obvious, but she insists that it is critical to keep asking, What are the trends? Where is the industry going? What's going on in Canada? "It's almost like you have this radar screen. You can't do that sitting behind your desk. You have to get out there and talk to people."

To Stephenson, entrepreneurship and innovation in the world's largest telecommunications company are not that different from a small startup. "You're just doing it on a slightly grander scale. The innovation certainly starts in our Bell Labs, in the R&D facility. And they are no different from entrepreneurs probably in a garage, except they have a few more resources at their disposal. We make the mistake by saying either you're an entrepreneur, or you work for a large company. Innovation is as important to large companies as it is to small ones, but sometimes is even harder to foster."

She also knows that while hindsight is 20/20, foresight is not. Stephenson knows it is easy to miss seeing a big change. "I remember, maybe 15 years ago, when the telephones were going to introduce voicemail. And I remember thinking, Who would ever want to pay $5 a month for voicemail, when they have an answering machine at home?" She thinks that mobile Internet will be like the voicemail of yesterday. A technology that is still new to many, in a few years it will be so pervasive we'll take it for granted. The recent dip in high-tech markets has not dampened her enthusiasm. "The possibilities are limitless. As far as human ingenuity will take us. I expect a revolution in the way we live, work, and play—and see telecom as an enabler. The next revolution is not about hardware; it's about technology enabling the world of applications. The bubble may have burst on the dot-coms, but not on the demand for useful information. Don't mix up the stock market with demand. We are seeing an explosion in Internet and mobility. And a huge push to get bandwidth."

Stephenson's unique perspective on the industry may in part be explained by the circuitous route she took getting into it, beginning with a degree in social work. "I'd been working part-time as a telephone operator for Bell Canada, and they were on a hiring program, hiring graduates into accelerated management development programs. The deal really was, if you could prove in a year that you would progress to mid-level management, then you got to stay—and if not, they fired you. Sink or swim. So that's how my career in telecom started." She worked through a wide range of positions at Bell over 25 years before reaching the top job at Stentor. "Fear of failure was my mentor, because I always thought, 'Oh, I'm in a new job; I don't know anything about it. I'm sure they're going to find out and fire me in the first two months.' And then of course that motivates you to learn everything that you can, and then you get accomplished, and then they move you again."

She makes no bones about the fact that the high-tech sector is still dominated by old boys. While the environment for women has improved, the barriers have not gone away and the glass ceiling is a reality. "The more power you've got, the more being a woman becomes an obstacle, although it becomes more subtle. And the more subtle it becomes, the more difficult it is to deal with." But if gender bias was a challenge, Stephenson rose to the occasion. "I also took it on as a bit of a mission. How are we going to get more women? How do we prevent people from hiring in their own image and diversify and that sort of thing? Although from time to time I ran into some situations which I wish hadn't occurred, it also gave me the strength and fortitude to say, I guess I need to do something about this and to do it a little more broadly, not just for myself."

From Stephenson's perspective, Canada is a great place to do business. "We have a highly educated workforce. We have a great infrastructure. Canada is also very safe and peaceful. The market in Canada is not that big, so you generally have much broader skills. For example, in Lucent in Canada, I have a much broader job than I would have if I were in the U.S. As a result, I have, I think, a much better business acumen. The disadvantage is that if you're going to grow, you have to grow outside your own home market, unlike the U.S. where the market is so huge that you can get to be a pretty hefty size without becoming international. But I think it plays to a lot of the strengths of Canadians. We are quite entrepreneurial. We are not the biggest and we aren't going to show up just because we're big, so as a result we use all those creative and innovative juices to find ways that we can compete with much larger companies."

NAVIGATING THROUGH THE WIND TUNNEL

INNOVATOR_ SANDRA WEAR, SERIAL ENTREPRENEUR, COACH INNOVATION_ CO-FOUNDED THE DOCSPACE COMPANY, AND SOLD IT TO CRITICAL PATH INC. ONLY 19 MONTHS AFTER LAUNCH FOR $568 MILLION (U.S.)

It's been over five years since Sandra Wear first dove into high tech and less than that since the company she co-founded, DocSpace Co., was sold to Critical Path Inc. for $568 million (U.S.). Still heady from the success of one of the biggest Internet infrastructure deals in Canadian history, Wear has found a new mission: "taking the craziness out of the startup and providing the sanity of a mature corporate entity." Her focus these days is on coaching other startups through the difficult growth stages; sharing the lessons she learned the hard way. She uses the phrase "wind tunnel" to describe the wildly turbulent transition phase that high-tech companies often face after they receive injections of investment capital.

Wear's new company, Tykra Inc., consults to tech businesses as they cycle from startup to mature corporation. There is no question they can use the help—nine out of 10 companies that get VC funding fail. In an industry where well-funded companies often crash from too much growth, too fast, Wear thinks there is a need for her service, especially given the dizzying rate of change in high tech.

> IN SOME RESPECTS THE "WIND TUNNEL PHASE" IS ANALOGOUS TO ADOLES-
> CENCE. WEAR'S APPROACH FOCUSES ON FOUR MAIN AREAS: BUSINESS (LEAD-
> ERSHIP AND MANAGEMENT); STRATEGY (INCLUDING MARKETING, BUSINESS
> DEVELOPMENT, AND FINANCE), CULTURE (INCLUDING POLITICS, COMMUNICA-
> TIONS, AND INNOVATION), AND ORGANIZATIONAL INFRASTRUCTURE (OR
> HUMAN RESOURCES). BY ADDRESSING THESE CORE ELEMENTS, SHE CAN
> HELP COMPANIES "AVOID BECOMING OVERWROUGHT IN OFFICE POLITICS,
> POOR MORALE, AND A DISINTEGRATING CULTURE THAT NEGATIVELY AFFECTS
> BOTTOM LINE."

She draws on her own experience with DocSpace as a record-breaking example of a company that moved successfully from launch, to hyper-growth, to acquisition, in just 19 months. During her tenure with DocSpace, Wear not only assisted in establishing corporate strategy, corporate growth plans, and major decisions (including investors, acquisition, and buy-out offers) but also helped to manage a team that grew from the initial five founders to a company with almost 100 employees. As the company matured, she says, she learned some hard lessons—especially while making the necessary shifts in corporate philosophy, decision-making, planning, and organizational structure. Now that she's earned her stripes in the trenches of high tech, Wear says she can help other startups improve their decision-making and put them on the path to profitability.

"Wind tunnels" are familiar to engineers as an environment in which to test the mettle of objects placed within them. Similarly, in high tech, the "wind tunnel" tests the capabilities of a company in its early stages. Startups are typically chaotic, whereas established firms are orderly. Startups tend to focus on short-term tactics for survival, while established firms are more strategic in their orientation. The transition from one stage to another can be traumatic. In some respects the "wind tunnel phase" is analogous to adolescence. Wear's approach focuses on four main areas: business (leadership and management); strategy (including marketing, business development, and finance), culture (including politics, communications, and innovation), and organizational infrastructure (or human resources). By addressing these core elements, she can help companies "avoid becoming overwrought in office politics, poor morale, and a disintegrating culture that negatively affects bottom line." She stresses the importance of "special attention to the role of leadership and management, the role of culture, establishing and designing corporate infrastructure, and the importance of a current and adaptable product marketing strategy."

According to Wear, the challenge exists as "you move from being founder-led, or entrepreneurially led, to management—that is, moving from focusing on merely developing a great technology to dealing with a lot of people issues." And this experienced management will most likely have to come from outside the company, which can upset the balance of the company and cause uncertainty as roles and positions are juggled. To make this likely scenario less traumatic, Wear believes it's important to start separating management and ownership early on and push the idea of being "replaceable." "Founders," she says, "are like parents; they will always be important, but you need to bring in experts to make your child fulfill his or her potential."

The parent/child relationship is an analogy Wear uses often to describe her role, particularly when describing the potential organizational challenges a company experiences during the growth phase. "Parenting styles change from child to adolescent to adult; likewise, so should the leadership, management, communication, and processes for an organization through its various life-stages." Often the qualities that make a startup so attractive (i.e., lack of hierarchy, no real processes, informal setting) are the very aspects that hinder a company's future success. Wear says, "The young agile startup can make decisions quickly, but hyper-growth demonstrates how the lack of systems decreases corporate effectiveness and efficiency."

In an effort to provide valuable support, Wear encourages her clients to address issues such as the timing and frequency of build cycles, hiring processes, reporting structure, cost-effective billing systems, and, most critically, communication issues. She often draws from her own experiences at DocSpace. "Even when our staff ballooned, we were constantly sharing information with the entire company via e-mail distribution meetings and so-called management meetings, but efficiency and effectiveness soon dropped. When you reach a certain size you have to move to a model that communicates 5 percent of the information to 100 percent of the company versus 90 percent of the information to 100 percent of the company, in order to stay streamlined."

Besides her efforts to share her entrepreneurial wisdom, Wear has an equally important mission—to "make more millionaires out of women in technology." She recently linked up with a U.S. nonprofit organization called Springboard, an initiative aimed at accelerating women's access to equity markets as both entrepreneurs and investors. Few would question the importance of Wear's efforts, since the high-tech industry continues to be overwhelmingly male-dominated. Despite the growing number of female entrepreneurs in Canada, when it comes to getting access to capital, women are often overlooked. According to Women Business Owners of Canada, women own one-third of Canadian businesses, and the number of women-led firms is increasing at twice the national average. Nevertheless, women are still almost twice as likely to be refused business loans as men. In addition, Wear says that currently only 6 percent of venture capital goes to companies headed by women.

Wear aims to level the playing field by creating forums for women to network and connect with some of the country's leading investors. Many women, she says, simply need the chance to become more familiar with talking about money and learning how to take the necessary leaps to access capital. By giving women the opportunity to gain confidence and provide awareness to a number of female-led and female-founded companies, Wear thinks that more women would not only be encouraged to compete as high-tech entrepreneurs but also to enter technology-related education and careers, thereby boosting the overall industry. As a result, she's setting up Springboard Canada events in Toronto and Vancouver to showcase top female-led companies.

Wear also admits that the lack of female role models in the high-tech arena represents another major obstacle to women's success there, and she hopes her success story will serve as an inspiration. Her own foray into high tech began in the marketing department of Vancouver-based Cyberian Network Corp., an Internet service provider. When the company opted out of producing a piece of software they owned, Wear and four partners launched DocSpace to develop the technology themselves.

As the only female founder in the group of fledging entrepreneurs, Wear says she brought a unique perspective to the startup dynamic. However, she says it was sometimes difficult to make her opinions heard and it was certainly a challenge as the only woman in the Toronto home/office where the group lived and worked together. Despite some of the drawbacks, Wear enthuses that overall, the hands-on experience in startup mode with mostly male colleagues was empowering and increased her "confidence and ability to succeed in a male-dominated culture."

Wear's impressive track record gives her credibility as a coach, whether she's helping business leaders hone their skills for hyper-growth, cultivating a crop of home-grown female entrepreneurs, or fostering a spirit of entrepreneurship within Canada. She's been there. She also has another advantage. She can also offer the perspective of a so-called "objective outsider," offering new approaches to what are often tightly knit teams of friends. Despite its challenges, her attachment to the startup environment is clear. "They're lots of fun. It's pretty crazy, intense, stressful and emotional but it's just such a great ride. ... You can create magic."[1]

---

[1] "Up-close: Sandra Wear, High Tech Nurturer." *Information Highways: Putting e-Content to Work*, August 2, 2001.

THINKING OPEN
SOURCE

INNOVATOR_ BOB YOUNG, SERIAL ENTREPRENEUR INNOVATION_ FOUNDED THE LARGEST
AND MOST RECOGNISED PROVIDER OF OPEN SOURCE SOFTWARE IN THE WORLD

Business schools teach students that a company needs to protect its source of competitive
advantage by "locking in" customers and building barriers to stop other firms from entering the market. Venture capitalists assess companies based on the value of their patents and
copyrighted intellectual property. Although it seems to defy established, century-old logic,
Red Hat actually sells software that you can get for free. The company doesn't hold any
patents. It hasn't copyrighted its intellectual property. According to the old rules, Red Hat
should have gone out of business long ago. But the company, founded in 1995 by
Canadian Bob Young and Mark Ewing, is thriving. What is the basis of their success? Two
words: open source.

Open source software means that anyone can access the source code. Why is this a good thing? Because *anyone* can improve the code, and people frequently do. Control shifts from the companies that develop proprietary solutions that don't interconnect to users who can build compatible systems based on accessible, open source standards.

Young explains. "In companies like Red Hat, we understand that the value of Linux is not actually Linux. It's not better, faster, cheaper software. It's actually about delivering control to your user, to your customer, because he never had control before. That's what makes Red Hat's customers so enthusiastic about what we're doing. And a lot

> "IF YOU TREAT EVERYONE, INCLUDING THE VAST MAJORITY OF THE WORLD WHO ARE ACTUALLY INDIFFERENT TO YOUR EXISTENCE, AS POTENTIAL ALLIES RATHER THAN POTENTIAL COMPETITORS, IT IS AMAZING THE AMOUNT OF SUPPORT, BOTH MONETARY, BUT MORE IMPORTANTLY NON-MONETARY, THAT YOU CAN GET OUT OF THE WORLD."

of the other little Linux companies are going out of business because they don't understand that this isn't about better, faster, cheaper software. It's actually about empowering your customer. Its why companies like IBM, who have caught on to the value that open source represents to the customers, are doing so well."

All that means that Red Hat is making money, a lot of money, by giving software away for free. You can download the files free of charge. You can copy and share them with your friends. The company's revenue comes mainly from services and support and from selling different versions of Linux on CD-ROM, some with additional software and varying levels of Web- or phone-based support. Prices for packages range from $29.95 to $2,500.

Originally developed in 1994, Linux is distributed in different versions by several companies, but Red Hat is the biggest (with over half the market). Young began, like many entrepreneurs, working from his home office, where he ran ACC—a Unix and Linux distributor. It was when he met Linux whiz Marc Ewing that he made the leap to the big time, combining his shrewd marketing talent with Ewing's technical genius. Young was named MC Marketer of the Year in 2000 for virtually redefining the rules of the game. Dan Kusnetzky, program director for operating environments and serverware at IDC, notes, "His company has demonstrated a genius in brand management and marketing that I don't think has ever been seen before in this business. They're doing what Starbucks did for coffee—selling an experience, not the product".[1]

Most people have used Linux and don't even know it, whether buying a book on Amazon.com or searching with Google. Linux-based computers were used to render the animation for the feature film *Shrek*. And the big players are taking notice—IBM has dedicated a staggering $1 billion (U.S.) to developing the platform, and Oracle will base its internal systems on Linux. Companies using Linux say it has a big impact on costs and efficiency. It is inexpensive and easy to maintain. And that is translating into a modest but growing share of the estimated $9.5 billion (U.S.) operating system market.

But Red Hat also inspires a kind of loyalty not often seen in an industry that changes at the speed of light. Recently it was named *Datamation* magazine's Product of the Year in the Network and Systems Software category (dislodging incumbent Microsoft): "While Red Hat and Linux have been around since the mid-'90s, it has taken time for the open source system to gain widespread trust among enterprise IT pros. If the votes of *Datamation* readers—primarily IT executives with experience and buying power—are any indication, Linux software appears to have gained a permanent and growing role in the enterprise."

While Microsoft may be winning the struggle for people's pockets, it also is the company many people love to hate. Red Hat unquestionably has the upper hand in the struggle for people's hearts. As Microsoft's principal nemesis, Young has been portrayed as an upstart David, struggling valiantly against the monster Goliath. But he does not convey

[1] Hassan Fattah, "Marketer of the year 2000: Robert Young," *MC Technology Marketing Intelligence*, February 2000.

the enmity one might expect in a warrior. Although his opponents have compared one aspect of Linux to a "cancer," Young quotes Gandhi in response: "First they ignore you, then they mock you, then they fight you, and then you win."

He insists that the objective is not to slay the giant. "Our job is not to make the life of 400 million people using personal computers a bit easier. Our job is to help the other 5.6 billion people on the planet."[2] Young does not see his principal focus as getting Microsoft users to switch, but rather winning the hearts and minds of new users. He recognised early on that he was not just marketing a product but the values and spirit that gave rise to Linux. It's an entirely new model.

He says, half-joking, "I'm restricted to being a visionary now. I'm not tough enough. I am too nice a guy." And his "nice guy" image pervades all aspects of his work. At Red Hat, Young applies the concept of open source to more than just his products. Open source products empower the users, and encourage users to work together to improve products. Open source management principles do the same thing, recognising the value of teamwork, and shifting from a top-down, management-directed environment into a collaborative one. This mindset also considers other companies potential allies rather than competitors. Young believes that many entrepreneurs make the mistake of competing, not co-operating. This too is contrary to conventional wisdom, and Young admits it could be construed as an example of that perceived Canadian weakness, niceness.

But it has worked for him throughout his career. "If you treat everyone, including the vast majority of the world who are actually indifferent to your existence, as potential allies rather than potential competitors, it is amazing the amount of support, both monetary, but more importantly non-monetary, that you can get out of the world."

This means that developing exclusive strategic alliances isn't such a good idea. "The problem with strategic alliances is the moment you're strategically allied with one player, you've by definition made competitors out of every other player in that industry. Instead, if you think in generic alliances, you build your business and your marketing model so that you can work with everyone."

---

[2] Tim Philips, "Fight the Power," *Director*, April 2002.

"At Red Hat, instead of slapping proprietary controls around our software and prohibiting people from downloading Red Hat and competing with us on it, we took exactly the opposite approach. We said it's open source. It allows our customers, and our users, and all the other Linux developers to help us improve our software. It reduces our cost of engineering. But more importantly, it eliminates us as a competitor to everyone. If you really don't want to give Red Hat money, you don't have to, you can just download it for free."

Although Young is convincing, the approach is at odds with everything we have come to expect in the hyper-competitive, dog-eat-dog high-tech environment. "At worst, our competitors thought we were naïve. But at best they would refer all their customers to us. If someone said, 'Which Linux should we use?' they would always say, 'Red Hat,' because they knew that we weren't locking their customers away."

Young's openness also applies to his approach to management and particularly to human resources. There's something else that "so many entrepreneurs get wrong that it just shocks me," Young says, talking about managing employees. Employees are an organization's "single most valuable asset," but many entrepreneurs forget this. An entrepreneur might start a company, but the employees are key to its ongoing success. "When I talk to small businesses and I realize there's one really, really smart guy running the business, it staggers me. Everyone else feels that their contribution is very marginal, because after all the really smart guy is making all the decisions. I don't care how smart the smart guy is, he's not as smart as the collective intellect of everyone working for him. If you just empower your employees, if you just treat them like human beings, and tell them on a regular basis you love them, and ask their advice, it is amazing how much smarter your company will be as a result of that than if you try and retain executive decision-making to yourself."

Certainly open source is a different approach to computing. Red Hat has a dramatically different approach to business. And Young has a different approach to management. But despite conventional wisdom, it may just be that nice guys don't always finish last.

LIES AN ENTREPRENEUR
TOLD ME DR. PAUL KEDROSKY

AFTERWORD

Most of what you read about entrepreneurship and innovation is a lie. Not an outright lie—nothing Enron-ian in its evil, or Blodget-esque in its mealy-mouthed duplicitousness—but a lie, nevertheless.

The trouble with the lie is that would-be entrepreneurs don't know what they're getting into, as you've seen in the profiles in this book. Before letting you in on the lie, however, let me explain my view of entrepreneurship and innovation. Because as a business professor, entrepreneur, sometime venture capitalist, and regular speaker on these subjects, I have many people tell me that they want to be entrepreneurs. I tell them that they don't want to be entrepreneurs. They say they do. I say they don't. And so it goes.

Why don't they? Well, for starters, entrepreneurs' hours are appalling. Think retail hours are bad? These schedules bring a whole new dimension of bad to the word "badness." Because entrepreneurs' hours are worse than retail. Way worse.

Consider: If you think there is any glamour in the CEO title, see how you feel after spending four days commuting up and down the U.S. seaboard (if this is Tuesday, this must be Jersey City) hawking your software to largely uninterested corporate buyers. And then heading to Las Vegas, Atlanta, or New York, doing the tradeshow junket, flogging QVC-style from dawn to late-night, telling bored attendees and punk analysts that your products are 20 percent better for 20 percent less money than the Other Product That They Know Better.

Indeed, you will learn such magnificent about-faces as the one you do when you find out that a much larger competitor has introduced a more or less identical product ("it validates our market"), but that is nowhere near enough to compensate for the time yanked out of your short life and left in airport check-in queues, taxis, and hotel rooms. Just try to live a normal life around this schedule: four flights a week, regular friskings, early morning marketing meetings, all-day technology strategy sessions, late-night customer dinners, and so little exercise as to be hardly worth the word.

So here is Lie #1: *Being an entrepreneur is fun.* It's not. It is hard work. Some people make it look easy and fun, but some people make being a fugu chef look easy and fun. And as anyone who has eaten poorly prepared fugu would tell you if they could, fun doesn't enter into it.

Most entrepreneurs fail. Statistics don't lie on this one, so trust me: Assuming you are thinking of creating a company, or maybe you already have a small one, your company will almost certainly fail. Present company in this book excluded, most entrepreneurs' businesses will assuredly go the way of Pets.com—if they're lucky. Why lucky? Because at least Pets.com had a chance: It raised money and got to rip around the block a few times funded by complete strangers' money. And there are few pastimes more fun than spending someone else's money with abandon.

But I digress. Because the danger with a book like this is that there is a sampling problem. You are talking to people who succeeded. Or at least to people who learned something useful. They make for great stories. Heartwarming and life-affirming stories. But the truth is, most entrepreneurs don't succeed—and most don't learn a darn thing from their flop. I still see business plans from entrepreneurs who haven't learned to do a search-and-replace on the phrase "business-to-consumer e-commerce."

Finance academics have heart attacks about this sort of thing; they call it a survivorship bias. That is a fancy way of saying that when you study only successes your results will be exceedingly biased. An analogy: Imagine Janus creates 20 mutual funds, eighteen of which do somewhere between lousy and middling, so Janus cans them; but two funds are standouts. Can we then reasonably use their performance as predictors of mutual funds in general? Of course not. But that's what we're doing when we just study successes.

So here is Lie #2: *Entrepreneurs succeed.* Most don't. 'Nuff said.

By the way, know what you should do if you want to impress a venture capitalist? Do what most entrepreneurs you read about here and elsewhere do: quit your day job, max out your credit card, and take a second mortgage on the house—in other words, do what you have to do to get your company going. Ready, fire, aim. That sort of thing. It is one of the founding notions of entrepreneurship, that the most successful entrepreneurs are risk-loving sorts, the kinds of people who take flyers, embrace chaos, and generally prefer living on the edge to living in the dull security of a normal job.

But that's wrong. As most anyone who's candid will tell you (and that set only intermittently intersects with that of venture capitalists) there are few dumber things than getting yourself deep into debt for a company—your own, or anyone else's. Why? Because debt messes you up. This happened recently to Bernie Ebbers (former CEO of near-defunct telco Worldcom) when he debt-financed purchases of oodles of company stock. A lot of debt will cross your eyes, make your stomach buzz, and generally cause you to spend way too much time worrying about what you can do to get the debt monkey off your back. And as Bernie will tell you, what is good for you and what is good for your company are not always the same thing when debt is involved.

What's more, outright quitting your job to make a *Business 2.0*-style splash isn't a very smart idea. Despite the garage-band stories, most entrepreneurs muddled their way into what they were doing while hanging into their normal jobs. Steve Jobs worked at Atari and HP while incubating Apple. Jim Clark taught undergraduates at Stanford while messing with the graphics technology that would become Silicon Graphics. The idea of quitting your day job to make your company work is a fatuous presumption foisted on naïve twenty-somethings by journalists, and by venture capitalist fat-cats soaking in 2.5 percent management fees from billion-dollar funds. Easy for them to say.

More extreme sorts, real fans of this risk-taking mentality, take an even stronger view. They see a willingness to lose your income and Go For It as a sign that you're really out there, that you're a mean mother risk-taker. It reminds me of one of my favorite Kafka short stories, "The Hunger Artist." In it a fellow tries to make a living starving to death in a zoo. For a while the crowds are huge, but eventually the novelty wears off and people discover the nifty black leopard next door. So much for the hunger artist's sacrifices. Entrepreneurial sacrifices are of the same order: They are interesting at first, and good story-fodder in retrospect, but that's about it.

So here is Lie #3: *You should quit your day job and start your company now.* A word of advice: Don't.

The next lie about entrepreneurship is that you need a lot of money. In particular, so this lie goes, you need a particular kind of money—immense amounts of venture capital (IAVC). The proof: an unholy number of business stories over the last decade started with the hook that Company X had received IAVC. The implicit assumption? That receiving IAVC was a little like being anointed by holy water, or economically equivalent to a purchase order from Wal-Mart—you had arrived.

But the truth is, as so many companies found out to their chagrin in the late 1990s, IAVC is not required. Matter of fact, IAVC is arguably a Bad Thing. It skews perceptions, confuses financial milestones with corporate ones, and generally makes life irritatingly complicated for companies just when they'd rather, all else being equal, be simplifying things, thanks very much. IAVC is a point mass, and like point masses in Einsteinian physics, it distorts space and time, making it difficult to remember just what was so important about getting things done so quickly, and about doing it out of a garage. Why bother, if you have IAVC?

And there is a deeper lie. Not only is IAVC a distorting force, plain old venture capital itself isn't necessary to run a successful business. Many businesses can be funded from cash flow, or from bank debt; venture capital (and the corresponding oversight) can be more irritant than assistant.

So here is Lie #4: *You need venture capital.* The truth? You don't.

I'm going to let you in on a secret: business schools are run wrong. The best thing most business schools could do is take the demographics of their incoming class, and apply those numbers to the population at large. Then ask people to apply to school, but throw all those applicants out of the admissions process. Now go back to your population profile and find people who were otherwise like those applicants, except, of course, that they didn't apply. Admit them. For free if necessary. They are your entrepreneurs, and likely to be wealthy alumni of the future. After all, they're too busy and involved to take two years off for accredited common sense. Those other folks? Bureaucrats.

I exaggerate—a little. But the truth is out there. You don't need an MBA to be a successful entrepreneur. While for a particular species of, ahem, entrepreneur, an MBA—in particular, the Harvard sort—became *de rigueur* for a spell, that was an unusual moment in time. Because most successful entrepreneurs don't have MBAs. Not only that, many of them (including the billionaire troika of Bill Gates, Steve Jobs, and Larry Ellison—and Andrew Carnegie, for that matter) haven't even completed undergraduate degrees. You can make a credible case that education, at least as currently constructed, gets in the way of being an entrepreneur. It slows you down.

Now, lest I set off a wail from college admissions offices and parents everywhere, I'm not suggesting that you drop out of school and wait for entrepreneurship to pick you up at the next bus stop. It doesn't work that way. Correlation, as they say, is not causation. Just because many entrepreneurs don't have MBAs, or even undergraduate degrees, doesn't mean that their lack of education was the cause of their success. Of course not, no more than the increasing numbers of car washes causes increasing murder rates in large cities (but there is a nice correlation!).

Successful entrepreneurs are in a hurry. They want to chase opportunities before the opportunity stops being an opportunity. And so they do a gut check (not head! not Excel!) and then they follow their muse, without worrying about silly things like accreditation. That is, at its core, what is so wonderful about entrepreneurship and innovation: it is anarchic and blind. It doesn't care who you are, what you eat, or where you come from, so long as your idea can separate customers from their money in volume.

So here is Lie #5: *You need a specific education to be an entrepreneur.* You don't. And anyone who says differently is likely an admissions officer, or knows one.

I could go on and on describing lies—and there are many of them—but I think that by now I have scared away the dilettantes, the people who were reading this book in *Inc.* fashion, looking for Five Easy Steps to be a successful entrepreneur, or Ten Businesses You Must Start Right Now! Here's hoping those people are gone, because now that they are, those of us who are left can have a productive conversation without all the noise.

Those of you who still remaining, all eleven of you, you're real entrepreneurs, real innovators. You know why? Because you won't be put off. You're persistent bastards. You know what you want. And while it's not widely known, the most powerful force in the universe is not gravity, or some esoteric bit of Hawking flotsam. No, the most powerful force in the universe is a person who knows what he or she wants, and will stop at nothing to get it.

Because most people don't know what they want. They may now and then, temporarily, but that is about it. Their convictions are fleeting, evanescent, swept away by time and change. True entrepreneurs and innovators aren't so easily shaken, because they have conviction. They know what they want, and they won't be stopped.

And far from worrying that they don't have all the data, they're happiest when they don't. Because they assume that if they don't, then no one else does, so they can rely on their instincts. You won't catch them noodling endlessly with some massively contingent spreadsheet doing contingency analysis—not that that's an altogether bad thing. Because real entrepreneurs thrive on ambiguity. They understand that they don't know everything.

And not only that, they understand that they will have to make (many!) decisions on imperfect information. They thrive on the idea. Analysis paralysis is for MBAs, not for real entrepreneurs. Because they live by the adage there are no bad decisions, other than no decision at all.

And those people who come up to me at conferences, the people who tell me that they want to be entrepreneurs but don't know how? Here is the truth: anyone who asks how to be an entrepreneur isn't one, by definition. If you think you need to ask someone else's permission, or you need to read a particular book, or take a particular course—or, heaven help you, take an MBA—then you don't have the Right Stuff to be an entrepreneur.

Here is some good news. Canada, by many measures, is wonderfully well prepared to lead the world in innovation and entrepreneurship. Our schools are among the best in the world, we outpace the U.S. in the production of scientists and engineers per capita, we have a world-beating lifestyle, we create companies at a rate comparable to the U.S., and (despite many claims to the contrary) we are in the top 10 globally in terms of venture capital as a percentage of GDP. Those are all good things.

So why isn't this book ten times larger? Why aren't there more Canadian entrepreneurs and innovators to highlight? It isn't the number of broadband connections to the home, nor is it the amount of risk capital in circulation. Instead, it has more to do with failure rate of small businesses.

One answer is that we simply don't have enough successes. While that might sound circular, it isn't, or at least isn't entirely. You need successful companies to breed successful companies. Partly because the role model is useful, but also for more practical reasons. You need the continual sloughing off of employees from these successful companies; they become the foot soldiers for the new revolution. You need the validation that comes from seeing other people succeed.

We saw some of the preceding in the dot-com euphoria, but people mostly learned the wrong lesson. Instead of learning that anyone can be an entrepreneur, people learned that anyone with a business plan and a line of patter could get venture funding and do an initial public offering. Not the same thing. And what's more, not true. Thankfully, however, those people are (mostly) gone, and so the rest of us can quietly go about our business.

And what is our business? It is filling swamps. Think about it. The business of innovating and being an entrepreneur is one of feeling your way along through uncertain terrain, through murky water containing unknown hazards. This economic swamp cannot be avoided, and with its soft bottom it is difficult to take soundings. What will market growth be? Where will our money come from? Who will buy? Who will sell to us? Where will we find good people? Are Macs really better than Wintel boxes?

To cross the entrepreneurial swamp and be real innovators, we have to answer questions. In other words, we have to fill the swamp. But as any Mafioso will tell you, it takes a lot of dead bodies to fill a swamp. Many companies fail before others succeed. Sometimes the first company succeeds, but not often. But here is the thing. Most entrepreneurs and true innovators don't care. They are not acting out of bizarre magnanimity when they toss themselves into the swamp. They genuinely think they'll float.

_Leonard Brody

_Wendy Cukier

_Ken Grant

_Matt Holland

_Catherine Middleton

_Denise Shortt

ABOUT THE AUTHORS

Leonard Brody is recognised as one of Canada's young entrepreneurial leaders. He is currently CEO of Ipreo, a new private equity research firm that assists the financial community in determining the success of public offerings. Previously, Leonard was VP of Corporate Development at Onvia.com (Canada's largest dot-com IPO) and jointly oversaw its Canadian operations. Leonard is also a member of the Board of Directors of the Information Technology Association of Canada and the Canadian E-Business Opportunities Roundtable.

Wendy Cukier has over 20 years experience as a consultant to industry and government. She is currently a professor of Information Technology Management in emerging technology trends and strategies at Ryerson University. She has presented and published over 100 papers and articles and is a regular contributor to *The Globe and Mail's* "Report on Business."

Ken Grant is the director of the new School of Information Technology at Ryerson University, where he is responsible for the establishment and supervision of the School's administrative and academic operations as well as teaching and research in business and technology strategy. Previous, Ken spent more then 20 years in high-tech management consulting as Vice President with A.T. Kearney and a portion with KPMG Consulting.

Matt Holland is a VP of The Boston Consulting Group and manages the Canadian practice. He has helped many major multinationals channel innovation into sustainable new business. Matt is the captain of the International Branding Team for the Canadian E-Business Opportunities Roundtable.

Catherine Middleton is coordinator of the e-Business Minor within the Faculty of Business at Ryerson University and an assistant professor in Ryerson's School of Information Technology Management.

Denise Shortt is a Harvard-educated writer and consultant. She recently joined the faculty at Ryerson University's School of Information Technology Management as a research associate. Denise is a contributor to the business anthology, *From the Trenches: Strategies from Industry Leaders on the New E-conomy* (John Wiley & Sons, 2001) and the co-author of *Technology with Curves: Women Reshaping the Digital Landscape* (HarperCollins 2000).